NATIONAL ACADEMIES Sciences Engineering Medicine

NATIONAL ACADEMIES PRESS
Washington, DC

Review of the U.S. Global Change Research Program's Draft Decadal Strategic Plan, 2022–2031

Committee to Advise the U.S. Global Change Research Program

Board on Atmospheric Sciences and Climate

Board on Environmental Change and Society

Division on Earth and Life Studies

Division of Behavioral and Social Sciences and Education

Consensus Study Report

THE NATIONAL ACADEMIES PRESS 500 Fifth Street, NW Washington, DC 20001

This activity was supported by contracts between the National Academy of Sciences and the National Aeronautics and Space Administration. Any opinions, findings, conclusions, or recommendations expressed in this publication do not necessarily reflect the views of any organization or agency that provided support for the project.

International Standard Book Number-13: 978-0-309-68994-6
International Standard Book Number-10: 0-309-68994-5
Digital Object Identifier: https://doi.org/10.17226/26608

This publication is available from the National Academies Press, 500 Fifth Street, NW, Keck 360, Washington, DC 20001; (800) 624-6242 or (202) 334-3313; http://www.nap.edu.

Copyright 2022 by the National Academy of Sciences. National Academies of Sciences, Engineering, and Medicine and National Academies Press and the graphical logos for each are all trademarks of the National Academy of Sciences. All rights reserved.

Printed in the United States of America.

Suggested citation: National Academies of Sciences, Engineering, and Medicine. 2022. *Review of the U.S. Global Change Research Program's Draft Decadal Strategic Plan, 2022-2031*. Washington, DC: The National Academies Press. https://doi.org/10.17226/26608.

The **National Academy of Sciences** was established in 1863 by an Act of Congress, signed by President Lincoln, as a private, nongovernmental institution to advise the nation on issues related to science and technology. Members are elected by their peers for outstanding contributions to research. Dr. Marcia McNutt is president.

The **National Academy of Engineering** was established in 1964 under the charter of the National Academy of Sciences to bring the practices of engineering to advising the nation. Members are elected by their peers for extraordinary contributions to engineering. Dr. John L. Anderson is president.

The **National Academy of Medicine** (formerly the Institute of Medicine) was established in 1970 under the charter of the National Academy of Sciences to advise the nation on medical and health issues. Members are elected by their peers for distinguished contributions to medicine and health. Dr. Victor J. Dzau is president.

The three Academies work together as the **National Academies of Sciences, Engineering, and Medicine** to provide independent, objective analysis and advice to the nation and conduct other activities to solve complex problems and inform public policy decisions. The National Academies also encourage education and research, recognize outstanding contributions to knowledge, and increase public understanding in matters of science, engineering, and medicine.

Learn more about the National Academies of Sciences, Engineering, and Medicine at **www.nationalacademies.org**.

Consensus Study Reports published by the National Academies of Sciences, Engineering, and Medicine document the evidence-based consensus on the study's statement of task by an authoring committee of experts. Reports typically include findings, conclusions, and recommendations based on information gathered by the committee and the committee's deliberations. Each report has been subjected to a rigorous and independent peer-review process and it represents the position of the National Academies on the statement of task.

Proceedings published by the National Academies of Sciences, Engineering, and Medicine chronicle the presentations and discussions at a workshop, symposium, or other event convened by the National Academies. The statements and opinions contained in proceedings are those of the participants and are not endorsed by other participants, the planning committee, or the National Academies.

Rapid Expert Consultations published by the National Academies of Sciences, Engineering, and Medicine are authored by subject-matter experts on narrowly focused topics that can be supported by a body of evidence. The discussions contained in rapid expert consultations are considered those of the authors and do not contain policy recommendations. Rapid expert consultations are reviewed by the institution before release.

For information about other products and activities of the National Academies, please visit www.nationalacademies.org/about/whatwedo.

COMMITTEE TO ADVISE THE U.S. GLOBAL CHANGE RESEARCH PROGRAM

JERRY M. MELILLO (*Chair*), Distinguished Scientist and Director Emeritus, Marine Biological Laboratory
KRISTIE L. EBI (*Vice Chair*), Professor of Environmental and Occupational Health Sciences and of Global Health, University of Washington
SUSAN ANENBERG, Associate Professor of Environmental and Occupational Health and of Global Health, George Washington University
SARA R. CURRAN, Professor of International Studies, Sociology, and Director of the Center for Studies in Demography and Ecology, University of Washington
PAUL FLEMING, Global Water Program Manager, Microsoft Corporation
SARAH K. FORTNER, Science Education Associate, Carleton College
MIRIAM GAY-ANTAKI, Assistant Professor of Geography and Environmental Studies, University of New Mexico
SHERRI W. GOODMAN, Senior Vice President, General Counsel, CNA Analysis & Solutions, Woodrow Wilson International Center for Scholars
ALISON M. GRANTHAM, Founder and Principal, Grow Well Consulting, LLC
KIMBERLY L. JONES, Associate Dean for Research and Graduate Education, Professor of Architecture, and Chair of Department of Civil and Environmental Engineering, Howard University
VALERIE KARPLUS, Associate Professor, Carnegie Mellon University
CARLOS E. MARTÍN, Rubenstein Fellow, Director of the Remodeling Futures Program, The Brookings Institution
LINDA O. MEARNS, Senior Scientist, National Center for Atmospheric Research
PHILIP W. MOTE, Professor, Vice Provost, and Dean of the Graduate School, Oregon State University
DEB A. NIEMEIER, Clark Distinguished Chair and Professor of Civil and Environmental Engineering, University of Maryland
OSVALDO E. SALA, Julie A Wrigley Professor of Life Sciences and Sustainability, Reagents' and Foundation Professor, Arizona State University
PAUL A. SANDIFER, Director, Center for Coastal Environmental and Human Health, College of Charleston
HENRY G. SCHWARTZ, JR., Independent Consultant
RACHAEL SHWOM, Associate Professor, Acting Director of Rutgers Energy Institute, Rutgers University
JOEL B. SMITH, Independent Consultant
ROBERT H. SOCOLOW, Professor Emeritus of High Meadows Environmental Institute, Princeton University
JULIE A. VANO, Research Director, Aspen Global Change Institute
ALYSSA K. WHITCRAFT, Associate Research Professor, University of Maryland
GABRIELLE WONG-PARODI, Assistant Professor, Center Fellow at the Stanford Woods Institute for the Environment, Stanford University
BRIAN L. ZUCKERMAN, Research Staff Member, Science and Technology Policy Institute

National Academies of Sciences, Engineering, and Medicine Staff
STEVEN STICHTER, Study Director
THOMAS THORNTON, Board Director of Board on Environmental Change and Society
AMANDA PURCELL, Senior Program Officer
LINDSAY MOLLER, Program Assistant

BOARD ON ATMOSPHERIC SCIENCES AND CLIMATE

MARY GLACKIN (*Chair*), Senior Vice President, The Weather Company, an IBM Business
CYNTHIA S. ATHERTON, Director of Science, Heising-Simons Foundation
CECILIA BITZ, Professor of Atmospheric Science, University of Washington
JOHN C. CHIANG, Faculty Scientist, University of California
BRAD R. COLMAN, Director of Weather Strategy, The Climate Corporation
BART E. CROES, Retired Research Division Chief, California Air Resources Board
ROBERT B. DUNBAR, William M. Keck Professor of Earth Science, Stanford University
EFI FOUFOULA-GEORGIOU, Distinguished Professor and Henry Samueli Endowed Chair at the Samueli School of Engineering, University of California
PETER C. FRUMHOFF, Director of Science and Policy and Chief Climate Scientist, Union of Concerned Scientists
VANDA GRUBIŠIC, Associate Director, National Center for Atmospheric Research
ROBERT KOPP, Director of the Rutgers Institute of Earth, Ocean and Atmospheric Sciences and Professor, Rutgers University
RUBY LEUNG, Battelle Fellow and Affiliate Scientist, Pacific North West National Laboratory and National Center for Atmospheric Research
ZHANQING LI, Professor in the Department of Atmospheric and Oceanic Science and the Earth System Science Interdisciplinary Center, University of Maryland
JONATHAN MARTIN, Professor and Chair of the Atmospheric and Ocean Sciences, University of Wisconsin-Madison
AMY MCGOVERN, Lloyd G. and Joyce Austin Presidential professor in the School of Computer Sciences, Oklahoma State University
JONATHAN A. PATZ, Director of the Global Health Institute and Professor and John P. Holton Chair of Health and the Environment, University of Wisconsin-Madison
J. MARSHALL SHEPHERD, Georgia Athletic Association Distinguished Professor and Director, University of Georgia
ALLISON STEINER, Associate Professor in the Department of Atmospheric, Oceanic and Space Sciences, University of Michigan
DAVID W. TITLEY, Professor of Practice in Meteorology, Director of the Center for Solutions to Weather & Climate Risk, Pennsylvania State University
ARADHNA TRIPATI, Assistant Professor of the Institute of the Environment & Sustainability, and the Director of the Center for Diverse Leadership in Science, University of California, Los Angeles Center for Policy Research
DUANE E. WALISER, Chief Scientist of the Earth and Technology Directorate, NASA Jet Propulsion Laboratory
ELKE U. WEBER, Gerhard R. Andlinger Professor for Energy and the Environment and Professor of Psychology & Public Affairs, Princeton University

National Academies of Sciences, Engineering, and Medicine Staff

AMANDA STAUDT, Senior Board Director
APURVA DAVE, Senior Program Officer
LAURIE GELLER, Senior Program Officer

APRIL MELVIN, Senior Program Officer
AMANDA PURCELL, Senior Program Officer
STEVEN STICHTER, Senior Program Officer
ALEX REICH, Program Officer
RACHEL SILVERN, Program Officer
PATRICIA RAZAFINDRAMBININA, Associate Program Officer
RITA GASKINS, Administrative Coordinator
BRIDGET MCGOVERN, Research Associate
AMY MITSUMORI, Research Associate
ROB GREENWAY, Program Associate
KYLE ALDRIDGE, Program Assistant
LINDSAY MOLLER, Program Assistant
SABAH RANA, Program Assistant

BOARD ON ENVIRONMENTAL CHANGE AND SOCIETY

KRISTIE L. EBI (*Chair*), Professor of Environmental and Occupational Health and of Global Health, University of Washington
BILAL M. AYYUB, Professor and Director of the Center for Technology and Systems Management, University of Maryland
LISA DILLING, Professor of Environmental Studies, University of Colorado Boulder
KENNETH GILLINGHAM, Professor of Economics, Yale University
LORI HUNTER, Professor of Sociology and Director of the Population Research Program, University of Colorado Boulder
KATHARINE L. JACOBS, Professor in the Department of Soil, Water and Environmental Science and the Director of the Center for Climate Adaptation Science and Solutions, University of Arizona
STEVEN H. LINDER, Professor and Vice Chair of the Department of Management, Policy and Community, University of Texas School of Public Health
GARY E. MACHLIS, Professor of Environmental Sustainability, Clemson University
MICHEAL A. MÉNDEZ, Assistant Professor of Environmental Policy and Planning, University of California, Irvine
ASEEM PRAKASH, Walker Family Professor for the College of Arts and Sciences and a Professor of Political Science, University of Washington
BENJAMIN K. SOVACOOL, Professor of Earth and Environment and Director of the Institute for Sustainable Energy, Boston University
MICHAEL VANDENBERGH, Professor of Law, Vanderbilt University School of Law
CATHY L. WHITLOCK, Professor of Earth Sciences and Fellow of the Montana Institute on Ecosystems, Montana State University

National Academies of Sciences, Engineering, and Medicine Staff

THOMAS THORNTON, Director
GRACE BETTS, Research Associate
SHARON BRITT, Program Coordinator
CHANDRA MIDDLETON, Program Officer
SITARA RAHIAB, Senior Program Assistant
JOHN BEN SOILEAU, Program Officer
HANNAH STEWART, Associate Program Officer

Acknowledgment of Reviewers

This Consensus Study Report was reviewed in draft form by individuals chosen for their diverse perspectives and technical expertise. The purpose of this independent review is to provide candid and critical comments that will assist the National Academies of Sciences, Engineering, and Medicine in making each published report as sound as possible and to ensure that it meets the institutional standards for quality, objectivity, evidence, and responsiveness to the study charge. The review comments and draft manuscript remain confidential to protect the integrity of the deliberative process.

We thank the following individuals for their review of this report:

THOMAS DIETZ, Michigan State University
SARAH HOBBIE, University of Minnesota
ELENA IRWIN, The Ohio State University
CHRISTINE KIRCHHOFF, University of Connecticut
KAI LEE, Owl of Minerva, LLC
BRIAN O'NEILL, Joint Global Change Research Institute
JONATHAN OVERPECK, University of Michigan
NANCY RABALAIS, Louisiana State University

Although the reviewers listed above provided many constructive comments and suggestions, they were not asked to endorse the conclusions or recommendations of this report nor did they see the final draft before its release. The review of this report was overseen by **WILLIAM C. CLARK,** Harvard University, and **ANTONIO J. BUSALACCHI,** University Corporation for Atmospheric Research. They were responsible for making certain that an independent examination of this report was carried out in accordance with the standards of the National Academies and that all review comments were carefully considered. Responsibility for the final content rests entirely with the authoring committee and the National Academies.

Preface

This report is the result of a review by the National Academies of Sciences, Engineering, and Medicine Committee to Advise the U.S. Global Change Research Program (USGCRP) of the USGCRP draft Decadal Strategic Plan (DSP) 2022-2031. The DSP 2022-2031 outlines priority areas of global change research at a time when our human and natural systems are increasingly experiencing impacts and when projections indicate the risks are expected to increase with additional change. The DSP recognizes that priority knowledge gaps shifted over the past decade as needs increased for useful and more inclusive data and information to ensure effective decision-making and implementation to increase resilience and sustainability. The report provides recommendations to strengthen the Plan and, in some cases, to expand strategic objectives.

The draft DSP embraces a systems-based perspective and a collaborative, inclusive approach based on four pillars (i.e., Advancing Science, Informing Decisions, Engaging the Nation, and Collaborating Internationally). The Committee recommends ways the language and tone of the draft DSP can be improved for clarity and to communicate a sense of urgency about the critical need for a comprehensive and robust U.S. Global Change Research Program to "assist the Nation and the world to understand, assess, predict, and respond to human-induced and natural processes of global change" (GCRA, 1990). This approach increases emphasis on the social sciences, community engagement (particularly of marginalized populations), and promotion of diversity, equity, inclusion, and justice in the production of science. The Committee provides guidance and recommendations for even deeper engagement with and connections between the public, decision-makers, and scientific communities. In this report, in both cross-cutting themes and discussions of individual DSP pillars, the Committee identifies content and strategy gaps—primarily missed opportunities for deeper integration across the plan.

The Committee to Advise the U.S. Global Change Research Program is the body within the National Academies of Sciences, Engineering, and Medicine responsible for advising USGCRP. We are indebted to the staff at the National Academies who provided guidance, input, and support throughout the writing of the report, particularly Steven Stichter, whose dedication and scientific understanding were critical throughout, and to Amanda Purcell, Dr. Amanda Staudt, and Dr. Thomas Thornton, whose deep technical knowledge and insights into the National Academies and USGCRP processes helped ensure an appropriately targeted report.

Jerry M. Melillo, *Chair*
Kristie L. Ebi, *Vice Chair*
Committee to Advise the U.S. Global Change Research Program

Contents

Summary   1

1 Introduction and Background   5
Statement of Task for the DSP Review, 6
Blueprint of This Review, 7

2 Cross-Cutting Themes and Issues to Strengthen the Draft Decadal Strategic Plan for 2022-2031   9
Terminology, 10
Incorporate a Stronger Sense of Urgency, 10
Strengthen Interconnections and Integration across the Plan, 12
Strengthen Coordination across Agencies, 13
Specify Key Research Outputs Where Additional Understanding over 2022-2031
   Could Advance Resilience and Sustainability, 16
Need for Strategic Flexibility over the Planning Period, 18

3 The Four Pillars   21
Pillars of the Draft Decadal Strategic Plan, 21
"Advancing Science" Pillar, 22
"Engaging the Nation" Pillar, 24
"Informing Decisions" Pillar, 27
"Collaborating Internationally" Pillar, 28

References   29

Appendixes

A Statement of Task for the Committee to Advise the U.S. Global Change Research Program   31

B USGCRP Transmission Memo   33

C Committee Member Biographical Sketches   37

D Line-by-Line Comments   49

Summary

The U.S. Global Change Research Program (USGCRP) is a federal government program mandated by Congress in 1990 to coordinate and integrate research and investments to "assist the Nation and the world to understand, assess, predict, and respond to human-induced and natural processes of global change" (GCRA, 1990).

The Global Change Research Act (GCRA) that established the USGCRP defined global change as "changes in the global environment (including alterations in climate, land productivity, oceans or other water resources, atmospheric chemistry, and ecological systems) that may alter the capacity of the Earth to sustain life." The GCRA identified a set of research elements to advance understanding of global change that include initiatives to understand the nature of and interactions among physical, chemical, biological, and social processes related to global change.

The GCRA requires development by USGCRP of a decadal strategic plan (DSP) and triennial updates to the strategic plan. The purpose of the strategic plan is to define "the goals and priorities for federal global change research which most effectively advance scientific understanding of global change and provide usable information on which to base policy decisions relating to global change" (GCRA, 1990, Sec. 104 (b) 1). USGCRP has developed its most recent draft strategic plan "for longer-term visioning for the Program and [encouraging] convergence among the agencies" (Appendix B).

The draft DSP for 2022-2031 reflects an important transition for the global change research enterprise, recognizing that priority knowledge gaps have shifted over the past decade as decision-makers increasingly grapple with simultaneously managing global changes across multiple sectors and communities. There are urgent needs for useful, more inclusive data and information to ensure effective and efficient decision-making and implementation and thus increase resilience to a rapidly changing environment. The draft DSP embraces a systems-based perspective and a collaborative, inclusive approach. This approach increases emphasis on the social sciences, community engagement (particularly of marginalized populations), and promotion of diversity, equity, inclusion, and justice in the production of science, policy, and action. These changes in approaches to global change research should help create new alliances and audiences for the DSP.

The Statement of Task for the Committee's review of the draft DSP 2022-2013 contains five questions (Appendix A). In considering the charge for this review, the Committee agreed that the scope and content of the draft DSP (question 1) is consistent with the GCRA and its strategic planning provisions (Section 104), recognizing that many of the directives in Section 104 were specific to the development of the first USGCRP Decadal Strategic Plan (GCRA, 1990). The bulk of this report and the Committee's recommendations consider and address questions 2-4, including the clarity, appropriateness, and fit of the *goals in the DSP* relative to the Nation's needs for understanding and responding to global changes, as well as opportunities for strengthening and expanding *coordination and integration* of global change research. In response to question 5, no major factual errors in the DSP were found.

The Committee applauds the work of the USGCRP in developing this draft Decadal Strategic Plan. In this review report, the Committee provides a series of recommendations on cross-cutting themes, and others specific to the DSP's four pillars, while recognizing the constraints within which USGCRP operates. The Committee hopes its recommendations contribute to an even stronger Decadal Strategic Plan for 2022-2031.

SUMMARY OF RECOMMENDATIONS

The core of the DSP is structured around four pillars (USGCRP, 2022): (1) Advancing Science, (2) Informing Decisions, (3) Engaging the Nation, and (4) Collaborating Internationally. The Committee's recommendations included those for individual pillars and cross-cutting recommendations that apply across the plan.

Recommendations on the Four Pillars

The draft DSP proposes four pillars for the next decade of USGCRP's work. The Committee supports these pillars and the advances they represent. The Committee also provides observations and recommendations to strengthen each pillar.

The draft DSP strengthens the role of engagement throughout the USGCRP's work. A structural change to the four pillars can reinforce the importance of engagement.

Recommendation: Reorder the sequence of the pillars to strengthen the interconnections between advancing science and engagement as Advancing Science, Engaging the Nation, Informing Decisions, Collaborating Internationally.

"Advancing Science" pillar: The Committee highlights opportunities to enhance the pillar by strengthening attention to urgency, interconnections, and outputs from global change research, as well as expanded indicators of global changes.

Recommendation: In the Advancing Science Pillar, (1) strengthen recognition of the urgency of global change issues, (2) define tangible outputs from this work, (3) make stronger connections to other pillars, and (4) increase the number and breadth of social and environmental indicators of global change, including for adaptation and resilience.

"Engaging the Nation" pillar: The draft DSP makes important space for new audiences to be engaged in the work of the USGCRP and global change research, including as partners in co-development of research and applications. Ongoing engagement and learning throughout the period of the DSP can strengthen these roles and contributions to global change work.

Recommendation: Include in the Engagement Pillar recognition of (1) new audiences for the DSP and mechanisms for engagement with them; (2) people- and place-based research to further deeper recognition of global change, associated risks, and effective and timely interventions; and (3) topics that would benefit from a sustained assessment process.

"Informing Decisions" pillar: The Committee recognizes the important ongoing work and commitments in the draft DSP for extending development and availability of climate information to support decision making; these information sources and platforms can serve as useful examples for addressing a broader range of global change challenges.

Recommendation: Expand on successful USGCRP efforts related to climate information products by providing specific outputs to assess progress and extend efforts to other global change issues.

"Collaborating Internationally" pillar: The Committee applauds the USGCRP for making international collaboration one of the pillars in the framework of the DSP for 2022-2031 and sees opportunities for expanding the types and focus of such collaborations.

Recommendation: Expand the discussion of international collaboration in the DSP to highlight examples of collaborations and emerging global change issues where U.S. or other national interventions could have international consequences and where international expertise could benefit the U.S. research enterprise to enhance resilience and sustainability nationally and globally.

Cross-Cutting Recommendations

Urgency: The Committee found that the draft DSP appropriately identifies the urgent nature of global change challenges and the importance of research to respond to those challenges. However, the draft DSP is uneven in identifying key global change challenges and desired outputs (research products). While a sense of urgency is conveyed for climate change, the sense of urgency is not equally well articulated for other global changes that affect the resilience of human and natural systems.

Recommendation: Maintain a strong sense of urgency throughout the DSP for meeting the challenges of global change for human and natural systems, including climate change, changes in land use and oceans, biodiversity, and the safety and security of food and water, among others.

Interconnections and Integration: The Committee recognizes the value of the four pillars as an organizing device for the DSP. However, pillars may act as silos, hampering interactions among the elements of the DSP. Highlighting cross-cutting themes and interactions among these themes may lead to a stronger strategic plan and more impactful research.

Recommendation: Stress interconnections and integration among pillars, including key themes and issues common to multiple pillars, and among global change issues, with enhanced integration of social sciences and systems-based research.

Coordination: The Committee finds that the DSP's discussion of international coordination in Pillar 4 is strong and commendable, but it notes that other types of coordination receive less attention across the other pillars. Given the USGCRP's mission to foster coordination across federal agencies, the DSP could do more to describe how the Program will further cooperation within the USGCRP and across other federal agencies to facilitate accomplishments under the final Decadal Strategic Plan ("final DSP").

Recommendation: Describe how USGCRP plans to strengthen coordination within, across, and beyond federal agencies within the "Advancing Science", "Engaging the Nation", and "Informing Decisions" Pillars, comparable to the level of specificity provided in the "Collaborating Internationally" Pillar.

Outputs: The draft DSP includes general statements about goals and research objectives, offering a direction of change. These statements and the final DSP would be strengthened by identifying ambitious research outputs that can be accomplished within the decadal time frame of the DSP, recognizing that budgets are uncertain.

Recommendation: Include illustrative examples of key research outputs in the DSP, where enhanced understanding of underlying science processes could advance policy and decision making on global change challenges to human and natural systems.

Strategic Flexibility: Ongoing global changes, along with changing vulnerabilities, capacities, and technologies, will continue to alter the context for global change research over the coming decade. The final DSP should explicitly aim to increase flexibility over the planning period to create new opportunities to increase resilience and sustainability at all levels, as well as insights from existing activities such as triennial reviews of the DSP. Regular evaluation of progress within and across all pillars of the final DSP would help create flexibility for mid-course corrections to enhance impact.

Recommendation: Add an approach to evolve the research questions, needs, and outputs in response to systematic evaluation and feedback from stakeholders and to respond to programmatic and technological developments.

1
Introduction and Background

The U.S. Global Change Research Program (USGCRP) is a federal government program mandated by Congress in 1990 to coordinate and integrate research and investments to "assist the Nation and the world to understand, assess, predict, and respond to human-induced and natural processes of global change."[1]

As detailed in the 2017 report *Accomplishments of the U.S. Global Change Research Program* (NASEM, 2017), two primary value-added activities of the USGCRP are *(1) strategic planning and coordination of global change research activities across the many federal agencies engaged in global change research and (2) high-level synthesis of global change research results and sharing them with decision makers and the American public.* These two activities have contributed to a variety of advancements in scientific capabilities, understanding, and applications. Examples highlighted in the 2017 report of scientific accomplishments enabled by the USGCRP collaborations included developing global Earth observing systems; improving Earth system modeling capabilities and understanding of carbon cycle processes; and increasing understanding of the multidimensional interactions between society, social dynamics, and global change, although progress in this area was noted to be uneven.

Decadal Strategic Planning

The Global Change Research Act (GCRA, 1990) requires development by USGCRP of a decadal strategic plan, as well as triennial updates to this plan. The purpose of the strategic plan is to define "the goals and priorities for federal global change research which most effectively advance scientific understanding of global change and provide usable information on which to base policy decisions relating to global change" (GCRA, 1990, Sec. 104 (b) 1). USGCRP developed the most recent DSP "for longer-term visioning for the Program and encouraging convergence among the agencies" (Appendix B).

Draft Decadal Strategic Plan, 2022-2031

In May 2022, the USGCRP released for public comment a draft decadal strategic plan for 2022-2031. This draft DSP was developed by a subgroup of the Subcommittee on Global Change Research (the SGCR, effectively the USGCRP "Board of Directors"). Critical inputs to the process included Dr. Jane Lubchenco's letter to Dr. J. Michael Kuperberg[2] (Lubchenco, 2021); comments and discussion with USGCRP interagency groups and at agency listening sessions, where many participants were from non-member agencies; the report of the Academies' Committee to Advise the USGCRP, *Global Change Research Needs and*

---

[1] In this report, the Committee understands and uses the term "global change" in the context of USGCRP's work to address global *environmental* changes, rather than non-Earth system drivers of global change, such as globalization. Consistent with the definition in the Global Change Research Act, the Committee uses "global change" to encompass climate change and other changes in the global environment, "that may alter the capacity of the Earth to sustain life" (GCRA, 1990).

[2] Dr. Lubchenco is the Deputy Director for Climate and Environment, Office of Science and Technology Policy (OSTP) in the Executive Office of the President. Dr. Kuperberg is the Executive Director of the USGCRP.

Opportunities for 2022-2031 (NASEM, 2021); and public comments on the prospectus for the DSP.

This draft DSP provides an overview of the USGCRP and its mission, vision, and structure for the coming decade. The core of the DSP is structured around four pillars (USGCRP, 2022):

1. **Advancing Science.** Advance scientific knowledge of interconnected natural and human systems and risks to society from global change.
2. **Informing Decisions.** Provide accessible, usable information to inform decisions on mitigation, adaptation, and resilience.
3. **Engaging the Nation.** Enhance the Nation's ability to understand and respond to global change by expanding participation in the federal research enterprise.
4. **Collaborating Internationally.** Build global capacity to respond to global change through international cooperation and collaboration.

The National Academies of Sciences, Engineering, and Medicine and the USGCRP

The National Academies has been an advisor to USGCRP planning efforts since the Program's formation (NASEM, 2017, Appendix D). In mid-2011, a new National Academies Committee to Advise the USGCRP (hereafter, the Committee) was formed and charged with providing a centralized source of ongoing whole-program advice to USGCRP (hereafter, the Program). The first major task of the Committee was to review a draft of the USGCRP 2012-2021 DSP. Now, a decade later, the Committee is tasked with reviewing the draft of USGCRP's 2022-2031 DSP.

STATEMENT OF TASK FOR THE DSP REVIEW

The Statement of Task for the Committee's review of the draft DSP 2022-2013 contains five questions (Appendix A):

1. Is the plan consistent with the direction provided in Section 104 of the Global Change Research Act?
2. Are the plan's goals clear and appropriate? Do they reflect the Nation's needs for information on climate and global change?
3. Does the plan show a clear strategy for coordination and integration that involves multiple disciplines and multiple agencies?
4. Does the plan communicate effectively with both the public and the scientific community?
5. Are there any factual errors, or major content areas missing from the plan that should be present if the Program is to achieve its overall vision and mission?

These five questions together with the transmittal letter from USGCRP to the Committee (Appendix B) guided the structure of our review.

In considering the charge for this review, the Committee agreed that the scope and content of the draft DSP (question 1) is consistent with the Global Change Research Act and its strategic planning provisions (Section 104), recognizing that many of the directives in Section

104 were specific to the development of the first USGCRP Decadal Strategic Plan (GCRA, 1990). The bulk of this report and the Committee's recommendations consider and address questions 2-4, including the clarity, appropriateness, and fit of the *goals in the DSP* relative to the Nation's needs for understanding and responding to global changes, as well as opportunities for strengthening and expanding *coordination and integration* of global change research. Finally, the Committee identifies editorial issues to be addressed (noted in the report and the appendix). In response to question 5, no major factual errors in the DSP were found.

BLUEPRINT OF THIS REVIEW

The Committee applauds the work of the USGCRP in developing this draft Decadal Strategic Plan and hopes its recommendations contribute to an even stronger Decadal Strategic Plan for 2022-2031.

This review is built around the draft DSP's four-pillar framework. The Committee's report begins by recognizing a set of themes and issues that weave together the four pillars and using these common themes and issues as a basis to make recommendations to strengthen the final DSP. Next, the review focuses on each of the four pillars and offers recommendations to refine and in some cases expand strategic objectives put forth by USGCRP. Finally, throughout this review, the Committee recommends ways the language and tone of the draft DSP can be improved for clarity and to communicate a sense of urgency about the critical need for a comprehensive and robust U.S. Global Change Research Program to "assist the Nation and the world to understand, assess, predict, and respond to human-induced and natural processes of global change" (GCRA, 1990).

In this review, the Committee commends the draft DSP for expanding the audiences for USGCRP's work by strengthening a commitment to engagement in global change research. Throughout this review, the Committee provides guidance and recommendations for even deeper engagement with and connections between the public, decision makers, and scientific communities. The Committee encourages dialogue between scientists and stakeholders consistent with the best practices of a true partnership to produce the useful and usable data, information, and tools needed to address global change challenges. In both cross-cutting themes and discussions of individual DSP pillars in this review, the Committee identifies content and strategy gaps—primarily missed opportunities for deeper integration across the plan.

2
Cross-Cutting Themes and Issues to Strengthen the Draft Decadal Strategic Plan for 2022-2031

A Strategic Plan to Enable Transformation

The Committee recognizes that the Program has catalyzed for the United States and the world transformative research that forms the foundation of our understanding of the global change systems and the extent of the associated challenges facing the nation and the globe. For the past three decades, climate change research was the primary focus of the federal agencies comprising the U.S. Global Change Research Program (USGCRP). This research played a critical role in advancing understanding of how human activities have altered the climate system and how these changes have increased impacts on human and natural systems.

The draft Decadal Strategic Plan (DSP) for 2022-2031 reflects an important transition for the global change research enterprise, recognizing that priority knowledge gaps shifted over the past decade as decision makers moved from questioning the extent to which recent climate change was caused by human activities to seeking evidence-based approaches to manage the increasingly severe impacts of climate change on multiple sectors and communities. There are urgent needs for useful, more accessible, and inclusive data and information that will ensure effective and efficient decision making and implementation to increase resilience in a rapidly changing environment. There are also urgent needs for continued research to identify processes and discoveries to enhance our understanding of global change in support of human and natural systems.

In this context, the draft DSP for 2022-2031 includes important advances in:

- Accelerating systems-based research that integrates natural and social sciences to understand climate and global change risks, responses, and societal needs at scales relevant to decision-makers;
- Expanding participation in the federal research enterprise, both within and across an expanded set of federal agencies, and with stakeholder communities;
- Strengthening efforts to engage decision-makers and users in targeting the Program's research and science products to meet the needs of different decision-making processes and contexts;
- Increasing emphasis on community engagement (particularly of marginalized populations), and situating diversity, equity, inclusion, and justice at the core of approaches and priorities;
- Emphasizing new research on unknowns including extreme events, attribution, and tipping points;
- Incorporating monitoring and evaluation, including of institutional structures; and

- Increasing cooperation with international organizations, initiatives, and research networks to further enhance the ability of the Nation and world to understand, assess, predict, and respond to global change.

Noting these advances, the Committee makes recommendations to further strengthen the DSP while recognizing the constraints within which USGCRP operates.

TERMINOLOGY

Global change: In the Committee's review of the Decadal Strategic Plan, consistent with prior reports (e.g., NASEM, 2021), we have adopted the GCRA's broad definition of "global change"; that is, "changes in the global environment that may alter the capacity of the Earth to sustain life." The GCRA provided examples of global change that included "alterations in climate, land productivity, oceans or other water resources, atmospheric chemistry, and ecological systems." Over the three decades since the GCRA became law, the list of global changes has grown to include changes in key ecological system attributes such as biodiversity. Biodiversity loss can affect the Earth's capacity to sustain life. While important research on a number of these and other global change phenomena has been the focus of individual USGCRP agencies (e.g., changes in land productivity by USDA), the Committee encourages the USGCRP to continue to pursue coordinated initiatives to understand the nature of and interactions among physical, chemical, biological, and social processes related to global change. These coordinated efforts on the interactions among global change phenomena have the potential for being important for understanding thresholds and tipping points in physical and social systems.

Human and Natural Systems: The Committee recognizes that there are multiple, overlapping phrases used to convey the interactions between human and natural systems. To simplify the discussion in this report and increase consistency with the language used by USGCRP, the phrase "human and natural systems" is used throughout this document. The Committee reinforces that these coupled systems encompass the entire biosphere and not just humans. Global change is affecting tightly connected human and natural systems where society is both affected by and driving global change. The Nation, through private and public institutions, determines rates of carbon emissions, the extent of protected lands, and rates of deforestation in the U.S. and abroad. Americans experience the impacts of climate change, land-use change, and biodiversity loss in the form of hurricanes, asthma epidemics, dust storms, reduced worker productivity, wildfires, destruction of property, and loss of lives, among others.

INCORPORATE A STRONGER SENSE OF URGENCY

The USGCRP was established because of a growing recognition that global changes such as climate change, as well as pollution, degradation of habitats, and biodiversity loss, were becoming widespread, and there was a need for more research into interventions to mitigate these problems (GCRA, 1990). Given the scale and scope of global change challenges, the "Fulfilling the Vision" section of the DSP[3] appropriately begins: "The urgent, transformative nature of global change requires a federal research enterprise equipped to meet the challenge."

---
[3] page 28, lines 2-8

The Committee agrees with this assessment of the urgency of risks associated with global change and the need for the USGCRP to pursue an ambitious agenda to identify research priorities and ultimately research outputs that, if achieved, would effectively and substantially reduce net greenhouse gas emissions, guide effective and timely strategies for greater adaptation and resilience to climate change, and address other global change challenges. The Committee further agrees with the critical importance of incorporating diversity, equity, inclusion, and justice challenges into global change research, including specific attention to low-wealth, minority, and marginalized communities, to ensure investments focus on promoting resilience, sustainability, and equity.

The draft DSP is uneven in identifying key global change challenges. While a sense of urgency is conveyed for climate change, the sense of urgency should also be well articulated for other global changes affecting the resilience of human and natural systems. Opportunities to better convey urgency throughout the DSP follow.

- The Committee urges the USGCRP to convey the urgency of the full spectrum of global change challenges in the Executive Summary and Introduction of the document, as well as in the concluding section.
- The discussion of urgency at the start of the "Fulfilling the Vision" section of the draft DSP includes clear and direct language that would strengthen the initial sections of the final DSP. Examples include:
    - Adding language from the "Fulfilling the Vision" section early in the Executive Summary.[4]
    - Editing the discussion of global change risks in the Introduction[5] to convey the same sense of urgency as expressed in the "Fulfilling the Vision" section. While the discussion mentions "increasingly disrupting Americans' lives and imposing high economic costs", that language does not convey the same sense of urgency.
- Should the USGCRP retain the current structure of the Executive Summary and Introduction in the final DSP, the Committee suggests that the existing IPCC and NASEM 2021 pull quotes[6] be placed prominently at the top of those sections to reinforce the urgency of these issues.

As the decade covered by the 2022-2031 DSP evolves, monitoring and evaluating the effectiveness of investments in research and implementation with respect to urgency can help ensure funding is targeted to the greatest challenges. As discussed below, this monitoring and evaluation will be more useful and effective if advances in research and implementation are measured against a clear set of outputs from the DSP.

Recommendation: Maintain a strong sense of urgency throughout the DSP for meeting the challenges of global change for human and natural systems, including climate change, changes in land use and oceans, biodiversity loss, and the safety and security of food and water, among others.

---

[4] i.e., page 2, before line 33
[5] page 6, lines 9-10
[6] page 4, lines 9-18

STRENGTHEN INTERCONNECTIONS AND INTEGRATION ACROSS THE PLAN

The Committee recognizes the value of the four pillars as an organizing framework for the DSP. However, pillars may act as silos hampering interactions amongst the elements of the DSP. For example, the Committee appreciates the value of the "Collaborating Internationally" pillar and recognizes that it must play a key role in the other three pillars, from generating new knowledge to communicating to a global audience. Similarly, assisting with decision making and generating new knowledge can be seen as part of a tight loop of supply and demand of global change knowledge. Highlighting cross-cutting themes and interactions among these themes may lead to a stronger strategic plan and more impactful research. Illustrations of interconnections and synergies across pillars are provided below as examples for consideration in strengthening cross-pillar integration in the final DSP.

Coordination among National Assessments

USGCRP is required under the GCRA to produce periodic assessments of the current state and trends for global change issues, known as the National Climate Assessment (NCA) (GCRA, 1990). The most recent of these, the Fourth National Climate Assessment, was released in 2018. In 2022, USGCRP was also charged with producing an "assessment of the condition of nature within the United States" (the "National Nature Assessment" [NNA]) (White House, 2022). Clear coordination between the NCA and NNA efforts would strengthen understanding of interconnections across nature and people and their joint impacts. Ensuring the assessments contain explicit points of contact where output from one assessment is designed as input to the other would increase the robustness of both assessments by highlighting the synergies and trade-offs across human and natural systems. The four-year periodicity of the reports could be complemented by special reports (in the spirit of the sustained assessment concept) responding to rapid changes in the state of the science and the environment as well as changes in societal perceptions and needs, similar to the special reports under the Intergovernmental Panel on Climate Change.

Integration of Social Science and Systems-based Research

The Committee welcomes the effort in the draft DSP to explicitly reference inter-disciplinary and trans-disciplinary research between social and natural systems. However, the broad range of social, behavioral, and economic sciences are not fully acknowledged and integrated into the draft DSP. Examples of approaches to emphasize the centrality of social sciences to global change research include:

1. Framing global change research around risks to people and nature, including through advancing research on implementation of adaptation and mitigation policies and programs, stocktaking of current efforts, increasing understanding of the process of adaptation, resilience stress testing, and integrated scenarios of global change and development pathways.[7]
2. Discussing how climate change interacts with other aspects of global change in complex ways, creating multi-hazard compounding and cascading risks that change over spatial and temporal scales, as discussed in the "Advancing Science" pillar of the draft DSP.

---

[7] See, for example, Nielson 2020 for a discussion of approaches for evaluating mitigation strategies for technical potential as well as likelihood of application by appropriate actors.

These risks can amplify impacts and alter constraints and barriers to effective interventions.
3. Including additional references to the use of scenarios across the pillars to strengthen research into their use for engaging decision makers and a general audience in planning for a range of plausible futures. Scenarios are an important tool for exploring the interactions of human and natural systems over temporal and spatial scales, yet their value is not captured in detail. They also are a tool for evaluating and communicating consequences of alternative actions. Scenario and story-telling approaches, co-developed with interested and impacted communities, can be especially effective in engaging people from a variety of communities, including those previously marginalized and/or excluded from most global change conversations.
4. Discussing how human responses and adaptations to global changes can both mitigate and exacerbate risks and vulnerabilities. These responses happen across different spatial, temporal, and organizational (e.g., from individuals to communities to institutions) scales, and are generated through interactions between human and natural systems.

To further strengthen the emphasis on systems-based research, the Committee suggests that the final DSP integrate insights from the social, behavioral, and economic sciences, as well as from disciplines that include the study of human culture, values, and ethics throughout, drawing on existing language from the "Social Sciences" section of the draft DSP.[8]

Recommendation: Stress interconnections and integration among pillars, including key themes and issues common to multiple pillars, and among global change issues, with enhanced integration of social sciences and systems-based research.

STRENGTHEN COORDINATION ACROSS AGENCIES

The Committee notes the draft DSP uses the concept of engagement in two senses. The first is engagement of federal agencies (and departments of agencies) in USGCRP. As noted in the 2016 report from the Committee on Enhancing Participation (NASEM, 2016), it "has become clear that the current group of member agencies is not adequate for addressing the breadth of the challenges that the United States faces" with respect to global change, and that "additional partnerships are needed to address all of the goals and objectives" described in the Program's strategic plan; that is more true of the 2022-2031 DSP than of the 2012 Strategic Plan.

The second type is broad engagement with researchers, practitioners, decision-makers, and affected communities through a process termed co-production, where appropriate. There are basic science questions for which co-production is neither appropriate nor informative. However, achieving the Program's goal of increasing the resilience of human systems to global change requires research to generate insights to inform decision-making. Co-production acknowledges that engaging with those most affected by decisions in the design and conduct of research will enhance uptake and effectiveness of possible solutions.

The Committee suggests strengthening the final DSP by clarifying these different types of engagement and what they mean for the Program over the next decade.

---

[8] e.g., incorporate the text on page 15, lines 29-33 into the introduction and move the bullet points of research topics (lines 36-45) across the other sections of the "Advancing Science" pillar.

The Committee commends USGCRP for elevating international collaboration to a pillar and for providing specific examples where research collaboration can yield near-term benefits (see discussion of the international pillar). The Committee notes less specificity with respect to coordination in the DSP's other pillars. Given the USGCRP's mission to foster coordination across federal agencies, the DSP could do more to describe how the Program will improve cooperation within the USGCRP and across other federal agencies to facilitate accomplishments under the final DSP. One commendable exception is the commitment to coordinating an interagency effort to make data from USGCRP research available, findable, usable, and customizable.[9] More examples like this would strengthen the final plan.

Additional opportunities for enhancing the DSP's discussion of coordination include the following four examples:

- *Within the USGCRP itself.* The Interagency Working Groups (IWGs) are mentioned in Box 2 of the DSP as part of the definition of the USGCRP, but the role of the IWGs in coordinating research across agencies on cross-cutting topics is not described, nor are the IWGs themselves called out in the Pillars, although they are alluded to in Box 5's discussion of health, food security, and carbon cycle assessment reports.[10] The Committee recognizes that it may not be feasible to focus on all the IWGs given space (and the fact that the IWGs change over time), but their role as USGCRP research-coordinating entities merits further mention. One approach might be to provide an example of the role played by an IWG in one of the Advancing Science areas (e.g., the Coasts Interagency Group as an example of cross-cutting, transdisciplinary research coordination related to the "Understand dynamics affecting the vulnerability of human–natural systems to global change impacts" section[11]).
- *Within the agencies that constitute the USGCRP.* In past National Academies reports (NASEM, 2016, 2021), the Committee encouraged the USGCRP to engage with federal departments and agencies that are not members of the Program. The Committee commends the USGCRP for incorporating engagement with these departments and agencies in the DSP. These reports also identified opportunities for USGCRP to enhance coordination between research and operating or regulatory sub-agencies *within* individual departments who can use USGCRP research products to help fulfill their missions (e.g., engaging the National Ocean Service and the National Marine Fisheries Service within NOAA with USGCRP research). The "Engaging the Nation" pillar discusses engagement with "non-USGCRP federal agencies and departments that need global change information to serve their constituencies."[12] The Committee suggests that the "Federal Agencies and Departments" section of this pillar also describe broader engagement within federal agencies and departments that have sub-agencies that "need global change information to serve their constituencies" but have not traditionally coordinated with the global change research community (NASEM, 2016; NASEM, 2021).
- *Coordination with other interagency efforts.* The USGCRP is not the only federal interagency group tasked with coordinating global change-related research. The NSTC Subcommittee on Ocean Science and Technology (SOST), the Fast Track Action

---

[9] page 18, lines 2-8
[10] page 18, lines 36-38
[11] page 15
[12] page 21, lines 23-24

Committee on Earth System Predictability Research and Development (ESP), the Interagency Arctic Research Policy Committee (IARPC), and the U.S. Group on Earth Observations Subcommittee (USGEO) have produced interagency research plans on topics that intersect with the mission of the USGCRP (SOST, 2018; ESP, 2020; IARPC, 2021; USGEO, 2019). Mentioning and discussing coordination with these other interagency groups can help clarify that USGCRP and its member agencies are involved with other critical entities coordinating global change research (e.g., SOST described ocean research priorities that USGCRP is leveraging as part of its coordinated global change research activities; USGEO is responsible for recommending the observation architecture upon which USGCRP agencies rely). Other potential coordination across interagency bodies can be a source of rich examples for a call-out box or in text. The SOST plan, for example, has a section titled "Understand a Changing Arctic" that calls out interactions with USGCRP and IARPC on Earth system models that integrate Arctic ice and atmospheric data more effectively (SOST, 2018). The Committee suggests a possible shift of the discussion of tipping points in the "Advancing Science"[13] pillar from the Antarctic to the Arctic, referring to coordinated research that spans the missions of USGCRP and IARPC.

- *With Federal boundary partners.* The previous USGCRP decadal strategic plan called for the creation of a set of "hubs" to engage in regional coordination activities (USGCRP, 2012). The inclusion of the RISAs, CASCs, and Climate Hubs in the "Engaging the Nation" pillar of the draft DSP reflects the creation and maturation of these organizations. The Committee suggests shifts in how these federal boundary partners are discussed in the final DSP. In previous reports (NASEM, 2021), these partners were identified as engaging in co-production activities and in working with stakeholders to design research in addition to their engagement efforts. These organizations could be referenced specifically, for example, in the "Enhance user engagement in the research design process" paragraph under the "Informing Decisions" pillar.[14] Given the importance of engagement, co-production, and user-focused research across all four pillars, the Committee suggests that these federal regional science organizations (and the sub-agencies that sponsor them) be actively integrated into the final DSP to improve the likelihood of realizing its objectives.

The Committee suggests that the discussions of engagement with external organizations and frontline communities as part of the "Engaging the Nation" pillar[15] and of co-production and user participation in the research design process in "Advancing Science" and "Informing Decisions"[16] also include research coordination efforts. The Committee welcomes the DSP's explicit involvement of users and other external stakeholder groups in research and research coordination processes, recognizing that doing so increases the research's usability and uptake into decision-making processes (Stern et al., 2021). However, the language used in these paragraphs of the DSP tends to assign USGCRP and its member agencies the primary role of reaching out to (and not engagement with) external stakeholders[17].

---

[13] page 10, lines 35-39
[14] page 19, lines 24-34
[15] pages 21-23
[16] pages 19, 23
[17] e.g., page 19, lines 32-33

Recommendation: Describe how USGCRP plans to strengthen coordination within, across, and beyond federal agencies within the "Advancing Science", "Engaging the Nation", and "Informing Decisions" Pillars, comparable to the level of specificity provided in the "Collaborating Internationally" Pillar.

SPECIFY KEY RESEARCH OUTPUTS WHERE ADDITIONAL UNDERSTANDING OVER 2022-2031 COULD ADVANCE RESILIENCE AND SUSTAINABILITY

The draft DSP includes statements about goals and research objectives that indicate a direction of change (e.g., "USGCRP agencies will continue to advance understanding of potential tipping points in the Earth system, emphasizing the complex interactions between physical and social systems that could cross thresholds and lead to tipping points."[18]). However, there is often little or no indication of what will be learned or developed and by when. The 2022-2031 DSP would be improved by providing examples of key research outputs.

The final DSP would be strengthened by identifying ambitious but achievable research outputs (e.g., essential improvements in science, in contrast to outcomes that identify how society will use these research outputs to improve human welfare and the environment) that can be accomplished within the decadal time frame of the final DSP, recognizing that budgets are uncertain. The Committee recognizes that it would be challenging to identify appropriate research outputs for all strategic plan research objectives, but instead welcomes identification of illustrative research outputs to strengthen the DSP. The Committee suggests that identification of any research outputs clearly states they assume at least level funding.

This suggestion is consistent with Dr. Jane Lubchenco's letter to Dr. J. Michael Kuperberg of May 18, 2021 (Lubchenco, 2021) that challenged the Program to address specific research objectives that would lead to a better understanding of tipping points, such as the loss of pollinating insects or changes in Atlantic circulation; identifying factors that limit natural sequestration of carbon; or global oceans and interactions between oceans, land, and air. The identification of outputs to address important national issues has been a concern for decades. For example, the National Academies Committee on Science, Engineering and Public Policy (COSEPUP) addressed the appropriateness of research goals in a 1999 study (NASEM, 1999). The report concluded, "…a full description of an agency's goals and results, which is a principal objective of GPRA, must contain an evaluation of research activities and their relevance to the agency's mission…. For applied research programs, agencies should measure progress" toward outputs. "For basic research programs, agencies should measure quality, relevance, and leadership."

A positive example for which the Committee commends USGCRP is the specificity of its description of the nature of collaborative activities that the Program will undertake to: (1) identify particular stakeholders and groups with which that collaboration will occur, (2) recognize that international collaboration is bi-directional and has the potential to increase U.S. capacity, (3) emphasize collaboration and capacity-building related to global change science in low and middle-income countries, and (4) prioritize involvement of under-represented groups and communities to facilitate collaborative research within the US and internationally. The degree to which these objectives are achieved over the life of the ten-year research plan can be

---

[18] page 10, lines 41-43

measured and assessed as the plan is implemented and when it is completed, although more specificity on research objectives, outputs, and measurements is warranted. As discussed later, this monitoring and evaluation can increase the flexibility of the plan during implementation, ensuring it addresses the urgent and immediate needs of the Nation.

Example Outputs

The Committee provides the following illustrative examples of what research outputs the DSP could include, but not necessarily for specific inclusion in the plan:

Example 1—Global changes drive harmful cyanobacteria blooms that pose risks to human and ecosystem health.

- Urgent science challenge—Quantify climate changes and nutrient pollution of fresh and brackish waters that stimulate cyanobacteria blooms that cause negative effects on human and ecosystem health.
- Approach—Expand the multi-agency (EPA, NASA, NOAA, and USGS) Cyanobacteria Assessment Network efforts to develop early warning indicators.
- Projected Output—A useful tool to help manage freshwater systems that are facing concurrent changes in drivers (i.e., warming and nutrient loading) of harmful cyanobacteria blooms.
- Link to policy and decision-making—Support for science-based management policies.

Example 2—The role of formal and informal education.

- Urgent science challenge—Advance research to better track and inform public understanding of the links between global change, human health and well-being, and inequities.
- Approach—Identify and scale up effective educational approaches that address climate and health equity outcomes such as K-12, informal education (e.g., museums), and community education led by frontline community organizations, faith-based organizations, extensions offices, and more.
- Projected Output—Strategies for professional development and training to build capacity among educators, practitioners, community members, emergency personnel, local government officials, and planners to increase equity and justice when implementing interventions to manage the health risks of climate change.
- Link to policy and decision making—Scaling up translation efforts (e.g., education, professional development, tool design, culturally-responsive interventions) that increase health and well-being, and reduce inequities.

The Committee recognizes that identifying outputs from the USGCRP is a challenging goal, but including objectives for research outputs that are ambitious but achievable would facilitate significant progress toward observing, understanding, and informing policies and decision making on key global change issues and would maintain the focus of the plan on the urgent needs of the Nation.

Recommendation: Include illustrative examples of key research outputs in the DSP, where enhanced understanding of underlying science processes could advance policy and decision making on global change challenges to human and natural systems.

NEED FOR STRATEGIC FLEXIBILITY OVER THE PLANNING PERIOD

Ongoing global changes, along with changing vulnerabilities, capacities, and technologies, will continue to alter the context for global change research over the coming decade. The final DSP should explicitly aim to increase flexibility over the planning period to capitalize on new opportunities to increase resilience and sustainability at all levels while maintaining focus on the most urgent needs of the Nation. New insights in one area of science may create opportunities to pivot portions of the strategic plan to rapidly advance knowledge and informed decision making. Similarly, recognition that investments are not resulting in hoped-for gains may suggest shifting funding to another scientific area where investment could lead to quicker advancements in understanding. Regular evaluation of progress within and across all pillars of the final DSP would help create flexibility for mid-course corrections.

Stakeholders can be a source of new thinking during implementation of the final DSP, bringing ideas to USGCRP and its member agencies that realign priorities and enhance collaboration and coordination. In finalizing the DSP, USGCRP may want to consider refining its language to encourage such stakeholder-driven activities. An example of a potential refinement relates to the "External organizations" section, "These efforts can support sustained engagement throughout the research-to-decision-support process."[19] Adding "and from the decision-support-to-research" would emphasize that opportunities also derive from decision-makers' feedback to researchers. Where appropriate, the Committee encourages refining language throughout the document to convey that the USGCRP and strategic plan are encouraging stakeholder-initiated engagement, research coordination, and participation. Further, it would be helpful for the final DSP to clarify how the federal agencies are organizing to receive such input.

New technologies will arise during implementation of the final plan, some of which could improve how to monitor global changes, build resilience, and reduce greenhouse gas emissions. Some of these technologies such as the Internet of Things (IoT) could facilitate compilation of massive data sets. Others, such as machine learning models, a form of artificial intelligence (AI), could play a valuable role in the analysis of big data. Machine learning has the capacity to discover patterns and trends buried within vast volumes of data that are not readily apparent to human analysts. Emerging visualization technologies, such as animations, also could have a role to play in USGCRP's strategies for engaging and informing the Nation. New technologies for energy generation and transmission, as well as for new battery storage, can accelerate greenhouse gas emission reductions. Along with the benefits of these new technologies come challenges, such as privacy and bias concerns, that must be considered and addressed for equitable applications of these methods, tools, and approaches.

---

[19] page 22, lines 7-8

Evaluation Strategy

The Committee applauds the USGCRP for including in the draft DSP a section on evaluation[20] and supports assessing how USGCRP products are being used. However, the goal is limited and does not address how the assessment would be carried out. The Committee notes that the USGCRP 2012-2021 plan (USGCRP, 2012) discussed using peer review, dialogue with users, and evaluation of participatory processes to assess progress, all of which continue to be relevant. The existing triennial reviews and updates of the DSP provide opportunities for incorporating evaluation findings to direct future work. More detail would be helpful on what aspects of the Program would be evaluated, recognizing the limited available space, and perhaps include peer review by a sample of users.

Recommendation: Add an approach to evolve the research questions, needs, and outputs in response to systematic evaluation and feedback from stakeholders, and to respond to programmatic and technological developments.

---

[20] page 20

3
The Four Pillars

PILLARS OF THE DRAFT DECADAL STRATEGIC PLAN

The draft Decadal Strategic Plan (DSP) proposes four pillars for the next decade of the U.S. Global Change Research Program's (USGCRP) work. The Committee supports these pillars and the advances they represent. The Committee also provides observations and recommendations to strengthen each pillar.

The draft DSP elevates stakeholder engagement as a critical input into its research agenda, emphasizing co-production of methods and tools to inform priorities and decisions. As such, engagement needs to be incorporated early into the strategic plan. While recognizing that the order of the pillars does not indicate either priority or sequencing between them, one way to demonstrate this critical role of engagement would be to reorder the pillars within the DSP.

Advancing Science should be the first pillar because it defines a primary focus of USGCRP, which is to enhance understanding of interconnected human and natural systems and risks to society from global change, and to identify effective approaches to increase resilience to global environmental changes. This science provides the foundation for the other pillars. As noted earlier, this pillar should more clearly articulate the urgency of advancing the science and the importance of clear goals and outputs for research and integration of engagement as appropriate to the research aim.

Engaging the Nation should be the second pillar in order to underscore USGCRP's recognition that expanding the impact of the federal research enterprise requires fostering meaningful engagement among scientists, affected communities, and decision makers. In a fundamental sense, the "Engaging the Nation" pillar functions as a bridge between "Advancing Science" and "Informing Decisions", by encouraging the engagement of a greater diversity of individuals in planning and conducting global change science and by fostering meaningful dialogue among scientists, affected communities, and decision makers.

The task of *Informing Decisions* on urgent issues for the Nation should be the subject of the third pillar. This pillar focuses on providing accessible, usable, and inclusive information to inform actions to advance mitigation, adaptation, and resilience.

Collaborating Internationally should remain as the fourth pillar, which underscores that humans are facing a set of interacting and compounding global problems that demand collaborative global research and solutions.

Recommendation: Reorder the sequence of the pillars to strengthen the interconnections between advancing science and engagement as Advancing Science, Engaging the Nation, Informing Decisions, Collaborating Internationally.

"ADVANCING SCIENCE" PILLAR

The Committee offers the following specific comments for the "Advancing Science" pillar to enhance the message of the need for urgent action for addressing global change challenges.

The research topics identified focus on general areas of research activity and data collection (e.g., "USGCRP agencies will continue to document biodiversity loss, global trends, and potential future losses due in marine, freshwater, and terrestrial ecosystems"[21]) but do not specify research outputs or identify new investments or research campaigns that will be undertaken during the DSP's timeframe. The Committee encourages USGCRP to move beyond documenting to investing in critical research to increase resilience and sustainability of human and natural systems.

The Committee further notes that the verbs associated with the research topics tend to use scientific language around increasing knowledge generally (e.g., "Improve understanding of the potential for abrupt, widespread changes in physical, natural, and human systems" in the Tipping Points section[22]). While scientific progress is uncertain and it is not realistic to assign specific dates to research breakthroughs, the descriptions of research activities do not convey to the public that these research efforts are timely, nor do they signal to academic researchers specific areas of short-term (or long-term) research focus by the USGCRP and its member agencies.

The cross-cutting recommendation to strengthen interconnections and integration also applies within individual pillars. Multiple, interacting global changes are affecting human and natural systems, requiring research that considers and synthesizes understanding of these multiple changes. For example, climate and land-use change can directly affect the rate of biodiversity loss. Moreover, biodiversity loss may affect the rate of climate change directly by affecting albedo and energy balance and indirectly by reducing the ability of ecosystems to respond to changes in climate. Similarly, soil erosion and salinization may affect albedo, climate change, and biodiversity loss. Advancing equity, for example, is a cross-cutting goal where defining outputs would enhance operationalization. Documenting disparities across communities could be an output that would help inform appropriate decisions and engage a diversity of communities.

In the "Advancing Science" pillar, each section would be strengthened by *stating the overall research goal* of the section (e.g., outcomes) and then *identifying specific research outputs* the Nation requires to understand, mitigate, and adapt to global change risks and, to the extent feasible, new investments to realize these outputs. The outputs should span USGCRP's focus on global change, including topics such as how research can help enhance the Nation's resilience and sustainability.

For example, the Committee considers the tipping points section introduction[23] to successfully identify the global change risk (irreversible changes that lead to significant societal impacts) and the research challenge that motivates its inclusion in the DSP (complex interactions between physical and social systems that may lead to tipping points, some of which may not be currently known or sufficiently characterized). For this section, an example of research outputs might be to focus on the five most consequential physical and human system potential tipping

---

[21] page 11, lines 11-12
[22] page 10, lines 32-33
[23] page 10, lines 33-45

points and characterize them sufficiently to avoid catastrophic surprises or to ensure sufficient warning to enable timely adaptation and mitigation efforts.

More specificity on how "Advancing Science" will relate to information products for "Informing Decisions" would increase the clarity of both sections and the final DSP. Improving decision making under uncertainty requires new data and knowledge to produce the information required by decision-makers for mitigation, adaptation, and resiliency efforts—and that information will need to support decisions made with imperfect knowledge of future events. But the discussion of individual scientific topics in "Advancing Science" does not identify what new information would be produced during the early years of the 2022-2031 plan period that will facilitate the development of the "Informing Decisions" products. Identifying which aspects of research in the "Advancing Science" pillar are needed to meet the goals of "Informing Decisions" would provide information to the USGCRP agencies and the research community regarding urgent topics to support decision-making efforts. Elevating and sharpening language around the "Advancing Science" topics related to decision making under uncertainty would be a welcome addition to the final DSP.

To make space in the DSP for addressing additional topics outlined in this report, Box 1 identifies topics in the draft DSP which may be considered for removal or de-emphasis in the final DSP.

Indicators

Currently, indicators are only mentioned in passing in the draft DSP.[24] However, in *Our Changing Planet 2021* (USGCRP, 2021), an activity on indicators was given as an example of a USGCRP achievement. Of the 18 global change indicators in the USGCRP's catalog,[25] all but three (Annual Greenhouse Gas index, Heating and Cooling Degree Days, and Billion Dollar Disasters) are physical science indicators rather than coupled human and natural systems or societal indicators. A recent update on activities of the Indicators Interagency Working Group (IndIWG)[26] highlighted their intent to identify and add new indicators to the USGCRP Indicator Platform, with an emphasis on social science and other non-physical science indicators.

To better address societal impacts and changes, USGCRP may consider balancing the 18 global change indicators by their relevance to societal impacts and responses by: (1) adding additional social indicators related to global change (e.g., health-related indicators or socioeconomic indicators related to key human systems, such as energy, food, water, health) and (2) relating the physical indicators to societal impacts (e.g., sea level rise to the status of coastal communities/systems in the U.S. facing increased flooding, subsidence, dislocation/relocation, etc.). An example from social science research is the demography of climate-driven migration. New indicators will likely emerge from time series observations in human and natural system research.

The selection of indicators identified for the Program should recognize and reflect their spanning and integration across the four pillars: Global change indicators result from science, are a mechanism for engagement and a tool for informing decisions, and should include enhancing and coordinating with international indicators.

---

[24] page 18
[25] See https://www.globalchange.gov/browse/indicators/catalog.
[26] See https://www.globalchange.gov/about/iwgs/indiwg.

> **BOX 1**
> **MAKING SPACE**
>
> The USGCRP requested that the Committee identify any components of the draft DSP that may be dropped or de-emphasized to help focus the priorities or make space for additions to the final plan. The Committee notes the following considerations for the "Advancing Science" pillar:
>
> - Remove the box describing transdisciplinary science. While (as noted previously in this report) this is an important aim, a box describing this approach is not needed in the DSP.
> - Figures 2 and 3 included in the draft DSP do not add sufficient value for the space they occupy and can be removed.
> - The Committee recommends a relative de-emphasizing of research on global climate sensitivity (see "Climate Sensitivity and Feedbacks" section) because it is not clear whether the relative insights on this topic will be greater than research in other important drivers of climate changes, such as understanding future climate variability (e.g., the frequency, persistence, and severity of extremes; regional feedbacks). While recognizing the importance of global climate sensitivity and key uncertainties, such as the role of "hot models," federal research should put relatively more resources into topics that can improve understanding of the drivers of climate variability (e.g., change in circulation patterns, or modes of inter-annual and decadal variability such as ENSO, PDO), their relationship to increased radiative forcing, and how climate variability and extreme events have changed and can continue to change.

Finally, USGCRP has a valuable role in improving the scientific underpinning of the indicators used from local to national levels to assess progress toward the UN Sustainable Development Goals.

Recommendation: In the Advancing Science Pillar, (1) strengthen recognition of the urgency of global change issues, (2) define tangible outputs from this work, (3) make stronger connections to other pillars, and (4) increase the number and breadth of social and environmental indicators of global change, including for adaptation and resilience.

"ENGAGING THE NATION" PILLAR

New Audiences for the Decadal Strategic Plan

The 13 federal agencies that constitute the USGCRP continue to serve the Nation by producing intramural research and funding extramural studies to better understand how global environmental change challenges will impact the U.S., primarily focusing on climate change (NASEM, 2017). The 2022-2031 DSP reflects an important transformation of the research enterprise. The DSP embraces a systems-based perspective with a collaborative, inclusive approach that places a strong emphasis on the social sciences and community involvement, particularly for vulnerable regions, while promoting diversity, equity, inclusion, and justice at each step. Doing so will serve new audiences for the DSP, within and outside federal agencies, such as the private sector.

The draft DSP is not always clear about the different types of audiences for research investments and their roles and responsibilities for conducting, translating, and using research results. Traditional audiences for earlier strategic plans included:

- *Federal agencies:* The DSP outlines the collective intent of the 13 federal agencies, signaling priority research directions within the USGCRP and across the federal government. This facilitates research coordination and collaboration across the federal agencies.
- *Congressional committees:* The DSP also signals to congressional committees the intent of budget requests.
- *The scientific enterprise investigating global environmental change*: This enterprise looks to the DSP to understand the strategic direction, to identify critical knowledge gaps that, if filled, would provide useful and usable input to decision making. The draft DSP is stronger when discussing research priorities in the physical science of global change than in the critical incorporation of social sciences and changes in natural systems.

The draft DSP takes the important step of prioritizing co-production of knowledge but offers few insights into what this progression means for the boundary partners that translate primary research into local action. Extramural funded research programs, such as RISAs, CASCs, Climate Hubs, and others, have established successful co-production models, but there is less intramural federal research that promotes co-production through traditional funding calls.[27] Acknowledging boundary partners as an important audience could accelerate the transition into co-production modes of research and implementation.

Engagement and Co-Production

The Committee commends the authors for responding to Dr. Lubchenco's charge to "accelerate action on two fronts" one of which is "ensuring that knowledge is understandable, accessible, and useable to the key stakeholders..." (Lubchenco, 2021). Only through expanding deep engagement and co-production, applied when appropriate, can knowledge stewarded by USGCRP activities achieve these goals. The Committee supports the DSP's emphasis on research that is place-based, people-first, and co-produced throughout the research-to-implementation pipeline; such research is particularly well suited to reduce inequities and reduce impacts on the most vulnerable.

However, the Committee also acknowledges the tension between people- and place-based research that is typically conducted as case studies at small scales, and the need to aggregate across case studies to understand lessons learned and best practices that can be scaled up in different locations. Ideally, such tensions should be identified and acknowledged in research planning and implementation. Too often the differences across a series of case studies are too large for syntheses and meta-analyses to be conducted. The Committee suggests the Program promote people- and placed-based research that provides not just knowledge for the scale of

---

[27] NSF's Convergence Accelerator is an example of a multi-disciplinary approach to bringing science to societal challenges; See https://beta.nsf.gov/funding/initiatives/convergence-accelerator.

local decision-making, but also intentionally provides usable insights that can be transferred to other locations and levels of decision-making.

The draft DSP is wise to call for "[enhancing] user engagement in the research design process,"[28] carried out appropriately. The Committee recognizes that engagement and co-production approaches need to match the requirements and implications of each research question. For example, participants engaged in setting priorities for fundamental research on deep ocean aspects of the carbon cycle will differ from those involved in setting research priorities for assessing the prospects for multiple uses of landscapes in particular places.

The Committee applauds the views of the draft DSP's emphasis on environmental justice, and notes that a particularly effective way to engage marginalized communities is through their existing trusted voices. Global change impacts can amplify the harms and impacts of structural racism and sexism on health and wellness outcomes. Research on coupled human and natural systems can center on ways to achieve equitable and just outcomes and reduce harm to marginalized individuals and communities through education, communication, and co-production of research and interventions to effectively shift organizational and governance practices.

When engaging tribal communities, whether federally recognized or not, the Program needs to honor the formal constraints of tribal sovereignty that shape U.S. federal engagement with tribes and understand the cultural and historical contexts, especially the history of injustice (including exploitative research practices) vis-a-vis tribes and tribal members.

Deepening the final DSP to strengthen research that centers on people and places impacted by global change can be achieved through the following approaches, as appropriate:

- *Support studies that investigate thresholds for "knowing," ways of knowing by different actors, and the ways in which certain types of evidence are used to justify action and inaction.* Such studies can inform context-specific decision support tools that quantify and forecast climate and other global change impacts; such decision support tools are not necessarily technology-based. Society needs to understand threats as well as adaptation and mitigation opportunities that are place-based, timely, and culturally/contextually appropriate.
- *Use best practices to develop tools to promote effective engagement,* (e.g., Hewitson et al, 2017) not just to "increase the capacity of ...users;"[29] the tools themselves must achieve high standards of usability. The Committee recommends explicitly stating that tools intended for users to access USGCRP information will achieve the highest standards of usability.
- *Establish inclusion and co-production as key precursors to analysis and options development addressing people and places affected by global change.* Differing priorities and needs within and across communities need to be identified; this requires engagement with communities impacted by global change and/or by the work of the USGCRP. Particular attention should be paid to ensuring that options do not exacerbate existing or historical social inequities.
- *Disaggregate studies and results to account for differential impacts.* Not everyone will be impacted in the same way by global changes. Differential impacts of climate change may have multiple sources, including historical structures and legacies of power.

---

[28] page 19 line 24
[29] page 18 lines 10-11

Consequently, global change studies should disaggregate results to help identify differential impacts.

Models for active co-development for climate and global change issues are available through existing regional partnerships throughout the country. Partnerships and centers, such as the NOAA RISAs, USGS CASCs, and USDA Climate Hubs, provide examples for regional and local co-development and sources for co-development literature. (See also NASEM, 2005; NASEM, 2008; NASEM, 2009; and NASEM, 2021.) Through initiatives such as these, federal agencies have existing experience and understanding of co-development approaches. Where agencies have well-established relationships with key stakeholders, they can share best practices with other agencies.

Sustained Assessment

One of the products of the Third U.S. National Climate Assessment (NCA3) was a new process for engaging government agencies, academia, the private sector, and civil society "to support their needs for usable, rigorous, and timely information and better connect science and decision-making." Referred to as the sustained assessment process, this collaborative process was crafted to foster "partnerships across a diverse and widely distributed set of non-governmental and governmental entities." A primary goal was "to produce timely, scientifically sound climate information products and processes" (all quotes: Buizer, 2016).

The USGCRP has previously used this approach to develop specialized reports to meet the needs of various stakeholders. In the draft DSP, there is reference[30] to conducting targeted assessments focused on and driven by the needs of Indigenous and Traditional Ecological Knowledge (ITEK) holders. It would be helpful for the DSP to briefly highlight other plans for the sustained assessment process.

Recommendation: Include in the Engagement Pillar recognition of (1) new audiences for the DSP and mechanisms for engagement with them; (2) people- and place-based research to further deeper recognition of global change, associated risks, and effective and timely interventions; and (3) topics that would benefit from a sustained assessment process.

"INFORMING DECISIONS" PILLAR

The Committee offers the following comments for the Informing Decisions pillar to communicate the urgency of action more effectively.

The Committee commends the USGCRP for identifying specific goals for the development of information products related to reaching net-zero emissions through carbon emission mitigation strategies, frequency of extreme events, geographically downscaled risk models (that incorporate effects on marginalized communities), and benefits and costs of adaptation and resilience actions.[31] Providing this technical basis for well-informed policy development efforts would be of great value to the Nation and internationally, helping to support robust decision making at all governmental and geographic scales. The Committee also commends the USGCRP for committing in 2022 to coordinating an interagency effort to make

---

[30] page 19
[31] page 17, lines 18-37

USGCRP-relevant data available, findable, usable, and customizable.[32] However, the goals of this activity are vague and do not include specific outputs that could be used to assess progress. Extending the draft DSP's efforts related to climate information products to other important global change issues would strengthen the final DSP.

The Committee also recommends careful consideration of the language in the "Informing Decisions" pillar to ensure that it is consistent with the draft DSP's commitment to engagement and co-development, and that it supports the bi-directional nature of such work.

Recommendation: Expand on successful USGCRP efforts related to climate information products by providing specific outputs to assess progress and extend efforts to other global change issues.

"COLLABORATING INTERNATIONALLY" PILLAR

The Committee applauds the USGCRP for making international collaboration one of the pillars in the framework of the DSP 2022-2031. Quantifying global change challenges and identifying effective interventions will require unprecedented efforts from the community of nations, big and small, developed and developing.

While the Committee notes that the examples of international coordination activities[33] given in the draft DSP tend to focus on collaborations around research synthesis, the Committee also suggests incorporating examples of international research collaborations including multi-nation-funded research and USGCRP-funded research conducted by and with international partners, especially those in low- and middle-income countries.

The Committee also encourages identification of emerging global change issues that will require international collaboration. For example, geoengineering (NASEM, 2021b) is being considered by some nations to mitigate climate change. It has complex human and natural system research components including issues of governance, where local interventions can have global impacts.

Recommendation: Expand the discussion of international collaboration in the DSP to highlight examples of collaborations and emerging global change issues where U.S. or other national interventions could have international consequences and where international expertise could benefit the U.S. research enterprise to enhance resilience and sustainability nationally and globally.

---

[32] page 18, lines 2-8
[33] e.g., page 22, lines 33-42

References

Buizer, J. L., K. Dow, M. E. Black, K. L. Jacobs, A. Waple, R. H. Moss, S. Moser, A. Luers, D. I. Gustafson, T. C. Richmond, S. L. Hays, and C. B. Field. 2016. Building a sustained climate assessment process. In *The US National Climate Assessment: Innovations in Science and Engagement*. K. Jacobs, S. Moser, and J. Buizer, eds. Cham: Springer International Publishing. https://doi.org/10.1007/978-3-319-41802-5_3.

ESP (Fast Track Action Committee on Earth System Predictability Research and Development). 2020. *Earth System Predictability Research and Development Strategic Framework and Roadmap*. Washington, DC: National Science and Technology Council. https://trumpwhitehouse.archives.gov/wp-content/uploads/2020/11/Earth-System-Predictability-Research-and-Development-Strategic-Framwork-and-Roadmap.pdf.

GCRA (Global Change Research Act). 1990. 15 U.S.C. § 2921.

Harley, A. G., and W. C. Clark. 2020. Capacity to Promote Transformations. In *Sustainability Science: A guide for researchers* (1st ed.). W. C. Clark and A. G. Harley, eds. https://doi.org/10.21428/f8d85a02.6bc9d7aa.

Hewitson, B., K. Waagsaether, J. Wohland, K. Kloppers, and T. Kara. 2017. Climate information websites: An evolving landscape. *WIREs Climate Change* 8(5):e470. https://doi.org/10.1002/wcc.470.

IARPC (Interagency Arctic Research Policy Committee of the National Science and Technology Council). 2021. *Artic Research Plan 2022-2026*. Washington, DC: Office of Science and Technology Policy. https://www.iarpccollaborations.org/uploads/cms/documents/final-arp-2022-2026-20211214.pdf.

Loorbach, D., N. Frantzeskaki, and F. Avelino. 2017. Sustainability transitions research: Transforming science and practice for societal change. *Annual Review of Environment and Resources* 42(1):599-626. https://doi.org/10.1146/annurev-environ-102014-021340.

Lubchenco, J. 2021. USGCRP Challenge Letter. May 18, 2021. https://www.whitehouse.gov/wp-content/uploads/2021/05/Lubchenco-to-Kuperberg-USGCRP-Challenge-Letter.pdf.

NASEM (National Academies of Sciences, Engineering, and Medicine). 1999. *Evaluating Federal Research Programs: Research and the Government Performance and Results Act*. Washington, DC: The National Academies Press. https://doi.org/10.17226/6416.

NASEM. 2005. *Knowledge-Action Systems for Seasonal to Interannual Climate Forecasting: Summary of a Workshop*. Washington, DC: The National Academies Press. https://doi.org/10.17226/11204.

NASEM. 2006. *Linking Knowledge with Action for Sustainable Development: The Role of Program Managers*. Washington, DC: The National Academies Press. https://doi.org/10.17226/11204.

NASEM. 2008. *Research and Networks for Decision Support in the NOAA Sectoral Applications Research Program*. Washington, DC: The National Academies Press. https://doi.org/10.17226/12015.

NASEM. 2009. *Informing Decisions in a Changing Climate*. Washington, DC: The National Academies Press. https://doi.org/10.17226/12626.

NASEM. 2016. *Enhancing Participation in the U.S. Global Change Research Program*. Washington, DC: The National Academies Press. https://doi.org/10.17226/21837.

NASEM. 2017. *Accomplishments of the U.S. Global Change Research Program*. Washington, DC: The National Academies Press. https://doi.org/10.17226/24670.

NASEM. 2021. *Global Change Research Needs and Opportunities for 2022-2031*. Washington, DC: The National Academies Press. https://doi.org/10.17226/26055.

NASEM 2021b. *Reflecting Sunlight: Recommendations for Solar Geoengineering Research and Research Governance*. Washington, DC: The National Academies Press. https://doi.org/10.17226/25762.

NASEM 2021c. *Progress, Challenges, and Opportunities for Sustainability: Proceedings of a Workshop–in Brief*. Washington, DC: The National Academies Press. https://doi.org/10.17226/26104.

Nielsen, K. S., P. C. Stern, T. Dietz, J. M. Gilligan, D. P. van Vuuren, M. J. Figueroa, C. Folke, W. Gwozdz, D. Ivanova, L. A. Reisch, M. P. Vandenbergh, K. S. Wolske, and R. Wood. 2020. Improving climate change mitigation analysis: A framework for examining feasibility. *One Earth* 3(3):325-36. https://doi.org/10.1016/j.oneear.2020.08.007.

Shonkoff, S. B., R. Morello-Frosch, M. Pastor, and J. Sadd. 2011. The climate gap: Environmental health and equity implications of climate change and mitigation policies in California—a review of the literature. *Climatic Change*, 109(1), 485-503. https://doi.org/10.1007/s10584-011-0310-7.

SOST (Subcommittee on Ocean Science and Technology). 2018. *Science and Technology for America's Oceans: A Decadal Vision.* Washington, DC: Office of Science and Technology Policy. https://www.noaa.gov/sites/default/files/2022-01/DecadalVision2018.pdf.

Stern, P. C., K. S. Wolske, and T. Dietz. 2021. Design principles for climate change decisions, *Current Opinion in Environmental Sustainability* 52:9-18. https://doi.org/10.1016/j.cosust.2021.05.002.

White House. 2022. Executive Order on Strengthening the Nation's Forest, Communities, and Local Economies. Washington DC. Office of the President. https://www.whitehouse.gov/briefing-room/presidential-actions/2022/04/22/executive-order-on-strengthening-the-nations-forests-communities-and-local-economies/.

Wilson, S. M., R. Richard, L. Joseph, and E. Williams. 2010. Climate Change, Environmental Justice, and Vulnerability: An Exploratory Spatial Analysis. *Environmental Justice* 3(1):13-19. https://doi.org/10.1089/env.2009.0035.

Woodruff, S. C., and M. Stults. 2016. Numerous strategies but limited implementation guidance in US local adaptation plans. *Nature Climate Change* 6(8):796-802. https://doi.org/10.1038/nclimate3012.

USGCRP (U.S. Global Change Research Program). 2012. *The National Global Change Research Plan 2012–2021: A Strategic Plan for the U.S. Global Change Research Program.* Washington, DC: National Coordination Office for the U.S. Global Change Research Program. https://downloads.globalchange.gov/strategic-plan/2012/usgcrp-strategic-plan-2012.pdf.

USGCRP. 2021. *Our Changing Planet: The U.S. Global Change Research Program for Fiscal Year 2021.* Washington, DC: USGCRP. https://doi.org/10.7930/ocpfy202.

USGCRP. 2022. *USGCRP Draft Decadal Strategic Plan.* Washington, DC: USGCRP.

USGEO (U.S. Group on Earth Observation Subcommittee). 2019. *2019 National Plan for Civil Earth Observations.* Washington, DC: Office of Science and Technology Policy. https://usgeo.gov/uploads/Natl-Plan-for-Civil-Earth-Obs.pdf.

Appendix A
Statement of Task for the Committee to Advise the U.S. Global Change Research Program

The Committee to Advise the U.S. Global Change Research Program (USGCRP or Program) provides ongoing and focused advice to the USGCRP by convening key thought leaders and decision makers at semiannual meetings, providing strategic advice, reviewing draft plans for the Program, and serving as a portal to relevant activities from across the National Academies of Sciences, Engineering, and Medicine. The Committee is broadly constituted to bring expertise in all the areas addressed by the USGCRP. The Committee will, over time, organize ongoing discussions, take on specific tasks, and possibly issue reports on a variety of issues of importance to the USGCRP and its major program elements.

1. In its role as a single entry point of contact to the National Academies, source of strategic input, and convener of strategic discussions with appropriate experts, the Committee to Advise the US Global Change Research Program will:
2. Provide ongoing, integrated advice to the USGCRP on broad, program-wide issues and documents when requested, including reviewing draft strategic plans and updates thereof.
3. Provide a forum for informal interaction between the USGCRP and the relevant scientific communities and other interested parties.
4. Provide a forum for exchange of experience and insights for integrating across science communities and disciplines, and improving linkages between officials of the Program and the science communities.
5. Improve the internal coordination across existing and future entities of the National Academies related to global change (including coordination across NAS, NAE, and NAM).
6. Help identify issues of importance for the global change research community. This implies a proactive role that goes beyond simply responding to requests from the USGCRP.
7. Interact with and provide advice to USGCRP relevant to its international activities, such as shaping the future of relevant international global environmental change programs.
8. In addition to producing National Academies' reports as tasked, the committee may help develop other work requests and promote collaboration on such efforts with appropriate units within the National Academies.

Statement of Task for the Review of the Draft USGCRP Decadal Strategic Plan

The Committee to Advise the U.S. Global Change Research Program (USGCRP) will conduct an independent review of the USGCRP's draft strategic plan concurrent with the public comment period. Per guidance to the USGCRP, the published decadal plan will be approximately 30 pages and written for a general audience. The review will address the following questions about the draft plan:

1. Is the plan consistent with the direction provided in Section 104 of the Global Change Research Act?
2. Are the plan's goals clear and appropriate? Do they reflect the Nation's needs for information on climate and global change?
3. Does the plan show a clear strategy for coordination and integration that involves multiple disciplines and multiple agencies?
4. Does the plan communicate effectively with both the public and the scientific community?
5. Are there any factual errors, or major content areas missing from the plan that should be present if the Program is to achieve its overall vision and mission?

Appendix B
USGCRP Transmission Memo

Date: May 20, 2022

From: Mike Kuperberg, Executive Director of the U. S. Global Change Research Program

To: The Committee to Advise the U.S. Global Change Research Program at the National Academies of Sciences, Engineering and Medicine

Topic: Advisory Committee Review of the U.S. Global Change Research Program's Decadal Strategic Plan (DSP) for 2022-2031

CC: Jane Lubchenco, Deputy Director for Climate and Environment at the White House Office of Science and Technology Policy and Co-Chair of the Committee on the Environment, National Science and Technology Council

Thank you for the time and expertise you bring to the Committee to Advise the U.S. Global Change Research Program (Advisory Committee or AC) as part of the National Academies of Sciences, Engineering and Medicine (NASEM). Your input is extremely valuable to the U.S. Global Change Research Program (USGCRP or Program) and to the White House Office of Science and Technology Policy. To advise USGCRP in finalizing its decadal strategic plan (DSP), NASEM has issued the following Statement of Task to the AC:

The Committee to Advise the U.S. Global Change Research Program (USGCRP) will conduct an independent review of the USGCRP's draft strategic plan concurrent with the public comment period. Per guidance to the USGCRP, the published decadal plan will be approximately 30 pages and written for a general audience. The review will address the following questions about the draft plan:

1. *Is the plan consistent with the direction provided in Section 104 of the Global Change Research Act?*
2. *Are the plan's goals clear and appropriate? Do they reflect the Nation's needs for information on climate and global change?*
3. *Does the plan show a clear strategy for coordination and integration that involves multiple disciplines and multiple agencies?*
4. *Does the plan communicate effectively with both the public and the scientific community?*
5. *Are there any factual errors, or major content areas missing from the plan that should be present if the Program is to achieve its overall vision and mission?*

Attached please find the draft Plan for your review; it was also released today for an eight-week public comment period. With this memo, we provide some background and context to accompany the Statement of Task for the review. We also look forward to meeting with you next week as you begin your work, with time for questions about anything in this memo and discussion with the DSP authors.

Overview
- The Global Change Research Act of 1990 (GCRA) directed USGCRP to develop a decadal strategic plan, including what should be included in that initial decadal plan. The GCRA also directed USGCRP to submit a revised plan at least once every three years thereafter.
- The GCRA's highest level guidance for the decadal plan is to "establish, for the 10-year period beginning in the year the Plan is submitted, the goals and priorities for Federal global change research which most effectively advance scientific understanding of global change and provide usable information on which to base policy decisions relating to global change."
- USGCRP has chosen to produce a decadal strategic plan on a regular basis, followed by triennial updates, as the development of a decadal plan allows for longer-term visioning for the Program and encourages convergence among the agencies.
- Our guidance for the current DSP is to write for a general audience, make it short (~ 30 pages) and high level, and cover a 10-year time horizon.
- The DSP reflects what the USGCRP collectively wants to do, but doesn't dictate to individual agencies in their planning or activities.
- The GCRA emphasizes USGCRP's role in research and in provision of information for use. USGCRP's work is thus meant to be policy-relevant but not policy-prescriptive.
- Annual implementation of the DSP is dependent on agency budgets.

Report Development
The DSP was developed by a subgroup of the Subcommittee on Global Change Research (the SGCR, effectively the USGCRP "Board of Directors"). This subgroup, the Executive Steering Committee, worked closely with the USGCRP Executive Director and the SGCR at each and every step. Critical input to the process included:
- The AC report on "Global Change Research Needs and Opportunities for 2022-2031."
- Public comments on the prospectus for the DSP (this is the first time a prospectus was made available for public comment).
- Comments and discussion with USGCRP Interagency Groups and at agency listening sessions, where many participants were from non-member agencies.
- Comments and discussion during the NASEM public engagement sessions on global change needs and risks in the areas of water, health, energy, food, and transportation/infrastructure.
- Comments from the SGCR, agencies, and Interagency Groups on the first order draft.
- Comments from the agencies, OSTP, and other White House components during clearance for public comment and NASEM review

How USGCRP Works
USGCRP is a Federal-only confederation, whose functions, including those below, are guided by laws, policies and long-standing practices.

- The Federal Advisory Committee Act created Federal advisory committees (FAC) as the mechanism for Federal entities to hold on-going interactions with non-Federal groups; per the GCRA, NASEM effectively functions as a FAC for USGCRP.
- The USGCRP helps coordinate implementation of agency budgets in the area of climate and global change, but does not coordinate, dictate, or evaluate agency budget submissions.
- USGCRP's scope (and thus, the scope of the DSP) aligns with the GCRA. USGCRP coordinates global change research, which is defined in Section 2 of the GCRA.
- USGCRP coordinates research across the agencies but doesn't commission or conduct it.
- USGCRP member agencies have a very wide range of missions and mandates; a coalition of the willing is required for any activities to move forward.
- Federal employees who participate in USGCRP inter-agency coordination activities typically do so voluntarily on top of their agency jobs; the personnel of the National Coordination Office are the only full-time USGCRP- dedicated employees in the Program.

Review Guidelines

Your review will be most helpful to USGCRP if:
- Suggestions for additional content areas or new language are accompanied by suggestions for deletions, to maintain total page length.
- Any major omission you note is accompanied by suggested high-level content that should be included.
- Any scientific errors that you flag are accompanied by suggestions on how to fix them.

Appendix C
Committee Member Biographical Sketches

Jerry M. Melillo (*Chair, NAS*) is a Distinguished Scientist at the Marine Biological Laboratory whose work focuses on understanding the impacts of human activities on the biogeochemistry of ecological systems using a combination of field studies and simulation modeling. His field studies include soil-warming experiments at the Harvard Forest in central Massachusetts and in northern Sweden and long-term observations of greenhouse gas emissions associated with deforestation in the Brazilian Amazon. Dr. Melillo and his team have developed and used a simulation model called the Terrestrial Ecosystem Model (TEM) to consider the impacts of various aspects of global change on the structure and function of terrestrial ecosystems across the globe. TEM is part of the Integrated Global Systems Model, an integrated assessment model based at the Massachusetts Institute of Technology. Dr. Melillo had a leadership role in the first three National Climate Assessments.

Kristie L. Ebi (*Vice Chair*) is a Professor in the Department of Global Health and in the Department of Environmental and Occupational Health Sciences, University of Washington. She has been conducting research and practice on the health risks of climate variability and change for nearly 25 years, focusing on understanding sources of vulnerability, estimating current and future health risks of climate change, designing adaptation policies and measures to reduce risks in multi-stressor environments, and estimating the health co-benefits of mitigation policies. She has supported multiple countries in Central America, Europe, Africa, Asia, and the Pacific in assessing their vulnerabilities and implementing adaptation policies and programs. She has been an author on multiple national and international climate change assessments, including the Fourth U.S. National Climate Assessment and the Intergovernmental Panel on Climate Change Special Report on Global Warming of 1.5°C. She is co-chair of the National Academies of Sciences, Engineering, and Medicine's Board on Environment and Society, the International Committee on New Integrated Climate Change Assessment Scenarios, and the Future Earth Health Knowledge Action Network. She is a member of the Earth Commission and of the Earth League. Dr. Ebi's scientific training includes an M.S. in toxicology and a Ph.D. and a Masters of Public Health in epidemiology, as well as postgraduate research at the London School of Hygiene and Tropical Medicine. She edited fours books on aspects of climate change and published more than 200 papers.

Susan Anenberg is an Associate Professor of Environmental and Occupational Health and of Global Health at the George Washington University Milken Institute School of Public Health. Dr. Anenberg studies the health implications of air pollution and climate change, from local to global scales. Dr. Anenberg has been a Co-Founder and Partner at Environmental Health Analytics, LLC, the Deputy Managing Director for Recommendations at the U.S. Chemical Safety Board, an environmental scientist at the U.S. Environmental Protection Agency, and a senior advisor for clean cookstove initiatives at the U.S. State Department. Her research has been published in top academic journals such as *Science*, *Nature*, and *Lancet Planetary Health*. She has also led or contributed to many science-policy reports on air quality and climate change published by U.S. EPA, the World Bank, the World Health Organization, the United Nations Environment Programme, and others. She has previously served on National Academies' planning committees for workshops on understanding air pollution from wildfires and leveraging remote geospatial technologies for precision environmental health. She also serves as the Secretary of the American Geophysical Union's GeoHealth section and as Editor of the GeoHealth journal. She received an M.S. and Ph.D. in Environmental Sciences and Engineering and Environmental Policy from the University of North Carolina and a B.A. in Biology and Environmental Science from Northwestern University.

Sara R. Curran is a demographer and Professor of International Studies, Sociology, and Public Policy & Governance at the University of Washington. She holds an Adjunct Professor of Global Health appointment at the University of Washington. She currently serves as the Director of the Center for Studies in Demography and Ecology and manages an NICHD research infrastructure grant to advance population science. Her research examines population dynamics (migration, settlement, and population change) in relation to climate and environmental change, as well as economic development. She received her Ph.D. in Sociology & Demography from the University of North Carolina at Chapel Hill in 1994.

Paul Fleming leads the Global Water Program for Microsoft. Paul joined Microsoft to build its corporate water stewardship program and has helped establish Microsoft as a leader in the corporate water stewardship space. In addition to driving the company's operational water commitments, Mr. Fleming drives collaborative partnerships with other companies and nongovernmental organizations and serves as the company's water subject matter expert, advising business groups on water issues. He is on the leadership committee of the Water Resilience Coalition, a group of 18 companies focused on collective action to improve conditions in water-stressed regions around the world, and serves on the steering committee of the CEO Water Mandate. Previously, Mr. Fleming developed and directed the Seattle Public Utilities' (SPU's) Climate Resiliency Group, where he was responsible for directing SPU's climate research initiatives, assessing climate risks, mainstreaming adaptation and mitigation strategies, and establishing collaborative partnerships. Mr. Fleming has been an active participant in several national and international efforts focused on water and climate change. He contributed to the 2014 U.S. National Climate Assessment, serving as a Convening Lead Author of the Water Resources chapter and the Sustained Assessment Special Report and a Lead Author of the Adaptation chapter. He is a Past Chair of the Water Utility Climate Alliance and chaired the Project Advisory Board of a research project focused on climate change and water management funded through the EU Horizon 2020 Program. Mr. Fleming has a B.A. in economics from Duke University and an M.B.A. from the University of Washington.

Sarah K. Fortner is currently a Science Education Associate at the Science Education Resource Center at Carleton College. She develops research and education collaborations to advance climate resilience and environmental justice. This includes projects engaging grassroots organizations, community development and health leaders, and faculty networks. Dr. Fortner leads civic visioning in the geosciences and higher education, including serving on the steering committee for the NAS Workshop on Service Learning in the Undergraduate Geosciences (2018) and advising the Association of American Colleges and Universities Civic Prompts in the Major (2020). Through collaboration with national organizations including the Union of Concerned

Scientists, the American Geosciences Institute, the National Association of Geoscience Teachers, and the American Geophysical Union she has led workshops and webinars helping faculty and interdisciplinary programs plan and strengthen sustainability education and local partnerships. Dr. Fortner also has 20 years of biogeochemical expertise including collaboration with the McMurdo Long Term Ecological Research Program. She also serves on the Academic and Research Council for the longest running glacier field education program, the Juneau Icefield Research Program. She holds a B.S. in Geology and Geophysics (1999) from the University of Wisconsin-Madison and a M.S. (2002) and Ph.D. (2009) in Earth Sciences from The Ohio State University.

Miriam Gay-Antaki is an Assistant Professor of Geography and Environmental Studies at the University of New Mexico. Their work focuses on human-environment relations in the era of anthropogenic global climate change. They trace climate change policy development ranging from formal political spaces, such as the UN Conference of the Parties, to scientific spaces, such as the Intergovernmental Panel on Climate Change, to the towns and communities where climate policies are implemented. Dr. Gay-Antaki's work investigates the participation, and sometimes the exclusion, of women scientists and stakeholders in international climate change research and policy arenas. For instance, their work on women climate scientists' perceptions of their participation in the Intergovernmental Panel on Climate Change, was published in the Proceedings of the National Academy of Science and covered in the media in outlets such as the *Popular Science* website and BBC Radio. This work influenced IPCC's decision to form a UN gender task force to increase equity in climate science. Dr. Gay-Antaki participated in this task force representing Mexico. In a publication in *Geoforum* they offer empirical and theoretical insights to better understand mechanisms that maintain social hierarchies in the climate debate and how these are resisted at the Conference of the Parties, the most important international meeting surrounding climate change policy. In the context of Mexico, Dr. Gay-Antaki studies the ways in which societal structures shape the development and implementation of transnational climate change policies such as gendered climate interventions. This work appears in the *Journal of Latin American Geography* and *Gender, Place and Culture.* To build healthy and resilient communities by facilitating dialogue among diverse interest groups is one of their priorities as the new Associate Director for the RH Mallory Center for Community Geography at the University of New Mexico. They have also worked alongside the Aspen Global Change Institute and 500 Women Scientists, the Mexican Council for Science and Technology Network on Gender, Environment and Society, and the National Autonomous University of Mexico Multidisciplinary Climate Change Team toward this goal. Dr. Gay-Antaki received their Ph.D. in Geography from the University of Arizona in 2017.

Sherri W. Goodman is an experienced leader and senior executive, lawyer, and board director in the fields of national security, climate change, energy, science, oceans, and environment. Ms. Goodman serves as the Secretary General of the International Military Council on Climate & Security, the global forum for military leaders and security professionals dedicated to addressing the security risks of a changing climate. She is a Senior Fellow at the Wilson Center's Polar Institute and Environmental Change & Security Program and Senior Strategist at the Center for Climate and Security. Previously, she served as the President and CEO of the Consortium for Ocean Leadership. Ms. Goodman served as Senior Vice President and General Counsel of CNA (Center for Naval Analyses) where she was also the founder and Executive Director of the CNA

Military Advisory Board. Ms. Goodman served as the first Deputy Undersecretary of Defense (Environmental Security) from 1993-2001. She has practiced law at Goodwin Procter, as both a litigator and environmental attorney, and has worked at RAND and SAIC. Ms. Goodman is a life member of the Council on Foreign Relations, served on its Arctic Task Force in 2016, and chaired the Advisory Committee on Governing Solar Geo-Engineering in 2022. A summa cum laude graduate of Amherst College, she has degrees from Harvard Law School and Harvard Kennedy School. She received an Honorary Doctorate in Humane Letters from Amherst College in 2018.

Alison M. Grantham is a scientist who brings a methodical, analytical, and quantitative approach to her work with nonprofits, private sector businesses, and foundations to improve our food system through her practice, Grow Well Consulting. Current and recent projects include improving climate impacts of pasture-raised poultry; greenhouse gas, water and waste footprinting for an indoor agriculture business; a national urban food waste and food insecurity analysis and report; global seafood traceability to support food safety and sustainability outcomes; and a FLAG sector scope 3 engagement for an international environmental NGO. Prior to launching Grow Well, she led Food Systems R&D and then Food Procurement at Blue Apron, overseeing food sourcing and procurement and implementing a national program to increase employee access to surplus product, as well as local communities through partnerships with Feeding America. While there, she also served on the National Academies' Ad Hoc Reducing Food Loss and Waste Committee. Previously, she led research at the Rodale Institute, including all aspects of organic and sustainable agriculture research. Alison holds a dual-title Ph.D. in Ecology and Biogeochemistry from Pennsylvania State University (2015) and B.A. summa cum laude in Biological Sciences and Environmental Studies from Mount Holyoke College (2008).

Kimberly L. Jones has over 26 years of experience in the civil and environmental engineering. She currently serves as Associate Dean for Research and Graduate Education (College of Engineering and Architecture) and Professor and Chair (Department of Civil and Environmental Engineering) at Howard University. Dr. Jones' areas of research expertise are in environmental justice, water quality and reuse, resource recovery, environmental management, and environmental nanotechnology. Dr. Jones holds a B.S. in Civil Engineering from Howard University (1990), a M.S. in Civil and Environmental Engineering from the University of Illinois (1992), and a Ph.D. in Environmental Engineering from The Johns Hopkins University (1996). Her research interests include water and wastewater quality, environmental policy, membrane separations, global water treatment, environmental justice, risk evaluation, and environmental nanotechnology. Dr. Jones has served on the Chartered Science Advisory Board of the U.S. EPA, where she chaired the Drinking Water Committee and was liaison to the National Drinking Water Advisory Council. She currently serves on the Advisory Committee for Environmental Research and Education at the National Science Foundation. She is an alternate Commissioner of the Interstate Commission on the Potomac River Basin in Washington, DC, where she chairs the committee on justice, equity, diversity and inclusion. She also serves on the Center Steering Committee of the Center for the Environmental Implications of Nanotechnology and on the Management Board of the Consortium for Risk Evaluation with Stakeholder Participation and as Associate Director for Diversity in the Urban Water Innovation Network. Dr. Jones has served on the Water Science and Technology Board of the National Academy of Sciences, and the

Board of Association of Environmental Engineering and Science Professors, where she was Secretary of the Board. She has served on several committees of the National Academies of Sciences, Engineering, and Medicine. She served as the Deputy Director of the Keck Center for Nanoscale Materials for Molecular Recognition at Howard University. Dr. Jones has received the Researcher of the Year award from Howard University, a Top Women in Science Award from the National Technical Association, the Outstanding Young Civil Engineer award from University of Illinois Department of Civil and Environmental Engineering, a NSF CAREER Award, an Outstanding Leadership and Service and Outstanding Faculty Mentor award from Howard University, and Top Women Achievers award from *Essence* Magazine.

Valerie J. Karplus is an Associate Professor in the department of Engineering and Public Policy at Carnegie Mellon University. Previously, Karplus served as an Assistant Professor of Global Economics and Management at the MIT Sloan School of Management. Karplus studies resource and environmental management in firms operating in diverse national and industry contexts, with a focus on the role of institutions and management practices in explaining performance. Karplus is an expert on China's energy system, including technology and business model innovation, energy system governance, and the management of air pollution and climate change. She works with a collaborative team of researchers to study the micro and macro determinants of clean energy transitions in emerging markets, with a focus on China and India. She teaches Entrepreneurship without Borders, New Models for Global Business, and is currently developing a new course, together with Professor Chris Warshaw in Political Science, on Global Energy Markets and Policy. She has previously worked in the development policy section of the German Federal Foreign Office in Berlin, Germany, as a Robert Bosch Foundation Fellow, and in the biotechnology industry in Beijing, China, as a Luce Scholar. From 2011 to 2015, she directed the MIT-Tsinghua China Energy and Climate Project, a five-year research effort focused on analyzing the design of energy and climate change policy in China and its domestic and global impacts. She is a faculty affiliate of the MIT Center for Energy and Environmental Policy Research, the MIT Energy Initiative, and the MIT Joint Program on the Science and Policy of Global Change. Karplus holds a B.S. in biochemistry and political science from Yale University and a Ph.D. in engineering systems from MIT.

Carlos E. Martín is a Rubenstein Fellow at the Brookings Institution's Metropolitan Policy Program and Director of the Remodeling Futures Program at Harvard University's Joint Center for Housing Studies. Martín, a trained architect and construction engineer, uses his technical training to connect the on the physical quality of housing and communities—technology, workers, and environmental performance and exposures—to its social outcomes. His areas of expertise include green housing, disaster mitigation, climate adaptation, housing quality, and building codes. Recent work includes evaluations of HUD's post-Sandy Rebuild by Design formation; the National Disaster Resilience Competition's Resilience Academies; home rebuilding rates with Community Development Block Grants for Disaster Recovery; and the Rockefeller Foundation's 100 Resilience Cities. Current independent research include studies of equity in energy-efficiency programs and flood mitigation infrastructure, planning and governance of adaptation authority, access to housing-related adaptation resources, and the capacity of climate-migrant receiving communities—the last supported by the National Academy of Sciences' Gulf Research Program. Previously, he was a senior fellow at the Urban Institute, assistant staff vice president for construction codes and standards at the National Association of

Home Builders, SRP Professor for Energy and the Environment at Arizona State University's Del E. Webb School of Construction and School of Architecture, and coordinator for the U.S. Department of Housing and Urban Development's Partnership for Advancing Technology in Housing. Martín received his B.S.A.D. in architecture from MIT and his M.Eng. and Ph.D. degrees in civil and environmental engineering from Stanford.

Linda O. Mearns is Head of the Regional Integrated Sciences Collective within the Computational and Information Systems Lab and the Research Applications Lab, and Senior Scientist, at the National Center for Atmospheric Research, Boulder, Colorado. She served as Director of the Institute for the Study of Society and Environment for 3 years ending in 2008. She holds a Ph.D. in geography/climatology from the University of California, Los Angeles. She has performed research and published mainly in the areas of climate change scenario formation, quantifying uncertainties, and climate change impacts on agroecosystems. She has particularly worked extensively with regional climate models. She has been an author in the Intergovernmental Panel on Climate Change's Climate Change 1995, 2001, 2007, 2014 and current (2021) Assessments regarding climate variability, impacts of climate change on agriculture, regional projections of climate change, climate scenarios, and uncertainty in future projections of climate change. For the Sixth Assessment Report, she is a lead author of the Atlas in Working Group I and a Review Editor for the North America Chapter in Working Group II. She led the multi-agency supported North American Regional Climate Change Assessment Program, which provided multiple high-resolution climate change scenarios for the North American impacts community and is currently the co-Chair of the NA-CORDEX regional modeling program. She has been a member of the National Research Council Climate Research Committee, the National Academy of Sciences (NAS) Panel on Adaptation of the America's Climate Choices Program, and the NAS Human Dimensions of Global Change Committee. She has worked extensively with resource managers (e.g., water resource managers and ecologists) to form climate change scenarios for use in adaptation planning.

Philip W. Mote is Vice Provost and Dean of the Graduate School and remains active in the Oregon Climate Change Research Institute (OCCRI). As Dean, he has established a number of strategic initiatives to accelerate student-centered and equitable graduate education including implementing holistic admissions, launching a new interdisciplinary program, training all graduate faculty in effective mentoring, and offering all graduate students opportunities to acquire transferrable skills. As Vice Provost, he has replaced the university's cumbersome approach to reviewing undergraduate and graduate programs with a holistic review of entire academic units. He served for several years in leadership of the 60,000-member American Geophysical Union: six years in leadership of the Global Environmental Change section, four years as member of the Council, two years as Vice Chair of the Council Leadership Team, and two years as a member of the Board of Directors. Dr. Mote was the founding director (2009-19) of OCCRI and remains involved in communicating climate science. He has served as a lead author for the Fourth and Fifth Assessment Reports of the Intergovernmental Panel on Climate Change, on three US National Climate Assessments, and on nine committees of the National Academies, including chair of the Review of the Climate Science Special Report.

Appendix C

Deb A. Niemeier (NAE) is the Clark Distinguished Chair in Energy and Sustainability at the University of Maryland, College Park, a professor in the Dept. of Civil and Environmental Engineering, and an affiliate professor in the College of Information Studies. Her current research includes collaborations with sociologists, planners, geographers, veterinary medicine, and education faculty to examine formal and informal governance processes in urban landscapes and to better characterize risk associated with outcomes in the intersection of finance, housing and infrastructure, and environmental hazards. Her international development work is aimed at agricultural sustainability, and her current education research is focused on data science in engineering and the operational challenges of K-12 infrastructure. She is a Fellow of the American Association for the Advancement of Science for "distinguished contributions to energy and environmental science study and policy development;" a Guggenheim Fellow for foundational work on pro bono service in engineering, and a member of the National Academy of Engineering and the American Philosophical Society.

Osvaldo E. Sala is the Julie A. Wrigley, Regents' and Foundation Professor at Arizona State University, where he contributes to both the School of Life Sciences and School of Sustainability. He is also the Founding Director of the Global Drylands Center. He came to ASU in 2010 from Brown University, where he was the founding Director of the Environmental Change Initiative and the Sloan Lindemann Professor of Biology. Dr. Sala has been trained as an ecologist working from the local to the global levels. He is known for his large-scale field manipulative experiments simulating climate change around the world. At the global scale, he has developed highly-cited scenarios of biodiversity change for the year 2100. His work has been truly interdisciplinary, collaborating with geologists, social scientists, mathematicians, and humanists. Dr. Sala received his Ph.D. (1982) and M.Sc. (1980) from Colorado State University and his B.Sc. from University of Buenos Aires (1973). Dr. Sala served in numerous international institutions, from the Millennium Ecosystem Assessment and the Intergovernmental Panel on Climate Change (IPCC) to President of the Ecological Society of America. He has been a contributor to several reports associated with global change including the IPCC's *Global Biodiversity Assessment* and the Millennium Ecosystem Assessment. His publications have been impactful as reflected in the more than 53,000 citations. He has received several recognitions to his academic work including being an elected Member of the American Academy of Arts and Sciences, the Academy of Sciences of Argentina, Fellow of the American Geophysical Union, the American Association for the Advancement of Science and the Ecological Society of America.

Paul A. Sandifer is Director of the Center for Coastal Environmental and Human Health at the College of Charleston, SC, and Deputy Director for the Center for Oceans and Human Health and Climate Change Interactions at the University of South Carolina. He is experienced in ecological and aquaculture research, natural resource management, science policy, and environmental health science. Previously he worked nearly 12 years in the National Oceanic and Atmospheric Administration (NOAA) overseeing the agency's Oceans and Human Health Program and as Senior Science Advisor to the NOAA Administrator and Chief Science Advisor for the National Ocean Service. Before NOAA, Dr. Sandifer worked 31 years as a scientist and manager, including as agency Director, with the South Carolina Department of Natural Resources. He served on the U.S. Commission on Ocean Policy, and he is an Honorary Life Member of the World Aquaculture Society and a Fellow of the American Association for the

Advancement of Science and the Ecological Society of America. He received a B.S. degree in biology from the College of Charleston and Ph.D. in marine science from the University of Virginia. His most recent work has concentrated on ocean health-human health linkages, human health impacts of disasters including the massive Deepwater Horizon oil spill, climate impacts in coastal areas, and ocean/science policy.

Henry G. Schwartz, Jr. (NAE) is a nationally recognized civil and environmental engineering leader who spent most of his career with Sverdrup Civil Inc. (now Jacobs Civil Inc.). In 1993 Dr. Schwartz was named president and chairman, directing the transportation, public works, and environmental activities of this international engineering firm before he retired in 2003. He has served on the advisory boards for Carnegie Mellon University, Washington University in St. Louis, and the University of Texas at Austin. He is President Emeritus of the American Society of Civil Engineers, the Water Environment Federation, and the Academy of Science of St. Louis, and the founding chairman of the Water Environment Research Foundation. Elected to the National Academy of Engineering in 1997, Dr. Schwartz has served on a number of National Research Council (NRC) study committees, including the Transportation Research Board's (TRB's) Committee for a Future Strategic Highway Research Program, and on the NRC Board on Infrastructure and the Constructed Environment. He chaired the policy study committee that produced the report *Potential Impacts of Climate Change on U.S. Transportation*. A convening lead author on National Climate Assessment (NCA) 2 and NCA 3, he has authored other papers focused on adaptation to climate change. For many years, he was on the TRB Executive Committee and served as Vice Chair of TRB's Subcommittee for NRC Oversight, in which capacity he was the final review authority for about 100 published, transportation research reports. Dr. Schwartz earned a Ph.D. from the California Institute of Technology and Master of Science and Bachelor of Science degrees from Washington University. He is a registered professional engineer.

Rachael Shwom is an associate professor in the School of Environmental and Biological Science's Department of Human Ecology and Acting Director of the Rutgers Energy Institute. She conducts research that links sociology, psychology, engineering, economics, and public policy to investigate how social and political factors influence society's responses to energy and climate problems. Rachael is currently a Co-PI on a multi-university, $3 million National Science Foundation grant on "Reducing Household Food, Energy and Water Consumption: A Quantitative Analysis of Interventions and Impacts of Conservation" and a newly awarded grant "Responses to Complex Disruptive Events: Cognition in a Socio-Political Context." She is Chair of the American Sociological Association's Environmental Sociology Section. Dr. Shwom was a Christine Mirzayan Science Technology and Policy Fellow at the National Academies of Sciences and a Michigan State University Environmental Science and Policy Fellowship recipient (Ph.D., Sociology 2009). From 2001-2004, Dr. Shwom worked in the utility demand side management sector and before that earned her M.E.M. from the Nicholas School at Duke University (2001) and B.A. in English and Textual Studies from Syracuse University (1999).

Joel B. Smith has been analyzing climate change impacts and adaptation issues for over three decades. He was a coordinating lead author or lead author on the on Third, Fourth and Fifth Assessment Reports of the Intergovernmental Panel on Climate Change. Mr. Smith was an author on three U.S. National Climate Change Assessments (NCA), including Chapter Lead on the International Chapter for the fourth NCA. He was a member of the National Academy of Sciences Panel on Adapting to the Impacts of Climate Change. Mr. Smith has provided technical advice, guidance, and training on assessing climate change impacts and adaptation to people around the world and to international organizations, the U.S. government, states, municipalities, and the non-profit and private sectors. He worked for the U.S. EPA from 1984 to 1992, where he was the deputy director of Climate Change Division. He has been a consultant since 1992, having worked for Hagler Bailly, Stratus Consulting, and Abt Associates. Mr. Smith received a B.A. from Williams College in 1979 (graduating magna cum laude), and a M.P.P. from the University of Michigan in 1982.

Robert H. Socolow is professor emeritus, Department of Mechanical and Aerospace Engineering, Princeton University; he taught a Freshman Seminar in the fall semester through 2021. Dr. Socolow earned his Ph.D. from Harvard University in theoretical high-energy physics in 1964, was an assistant professor of physics at Yale University from 1966 to 1971, and joined the Princeton University faculty in 1971 with the assignment of inventing interdisciplinary environmental research. Dr. Socolow is a member of the American Academy of Arts and Sciences, a fellow of the American Physical Society, and a fellow of the American Association for the Advancement of Science. His awards include the 2009 Frank Kreith Energy Award from the American Society of Mechanical Engineers and the 2005 Axelson Johnson Commemorative Lecture award from the Royal Academy of Engineering Sciences of Sweden (IVA). In 2003 he received the Leo Szilard Lectureship Award from the American Physical Society ("for leadership in establishing energy and environmental problems as legitimate research fields for physicists, and for demonstrating that these broadly defined problems can be addressed with the highest scientific standards"). Dr. Socolow is an associate of the National Research Council of the National Academies, in recognition of National Academies committee work. He served as a member of the Grand Challenges for Engineering Committee of the National Academy of Engineering and of the National Academies Committees on America's Climate Choices and America's Energy Future. Earlier committees included the Committee on Alternatives and Strategies for Future Hydrogen Production and Use (2002-2004) and the Committee on the Human Dimensions of Global Change (1992-98). From 2000 to 2019, Dr. Socolow and Steve Pacala were the co-principal investigator of Princeton's Carbon Mitigation Initiative, cmi.princeton.edu, a twenty-five-year (2001-2025) project supported by BP. His best-known paper, with Pacala, was in *Science* (2004): "Stabilization Wedges: Solving the Climate Problem for the Next 50 Years with Current Technologies." Dr. Socolow has also introduced "one billion high emitters," "committed emissions," and "destiny studies," as further conceptual decade-scale frameworks useful for climate change policy. His interests include energy efficiency in buildings, CO_2 capture and storage, technological "leapfrogging" by developing countries, and the conditionalities required for safe climate-change "solutions"— notably to protect against nuclear weapons proliferation and misuse of the land. Dr. Socolow was the editor of *Annual Review of Energy and the Environment*, 1992-2002. He was on the Board of the National Audubon Society, the Deutsche Bank Climate Change Advisory Board, and the Advisory Board of Lawrence Berkeley National Laboratory. He was the chair of the Panel on Public Affairs of the American Physical Society (APS), during which time he co-chaired the APS Technology Assessment, *Direct Air Capture of CO_2 with Chemicals* (2011). He is currently a member of the Science and Security Board of the Bulletin of the Atomic Scientists.

Julie A. Vano is the Research Director at the Aspen Global Change Institute, an organization dedicated to advancing global change science and solutions. Dr. Vano's research integrates elements of hydrology, water resource management, science policy, and climate impacts. She works closely with water utilities and U.S. federal water agencies to connect climate science and decision making. This has included being a lead in developing the Water Utility Climate Alliance's Leading Practices in Climate Adaptation report and the Bureau of Reclamation's Water Reliability in the West—2021 SECURE Water Act Report. She helped co-found the Mountain West Climate Services Partnership, an initiative to make science more relevant and accessible for communities across the Mountain West. Dr. Vano is president of the Science and Society section of the American Geophysical Union and holds an M.S. in Land Resources from the University of Wisconsin (2005) and a Ph.D. in Civil and Environmental Engineering from the University of Washington (2013). Dr. Vano's previous National Academies activities include work with the Water Science and Technology Board as a Christine Mirzayan Science and Technology Policy Fellow.

Alyssa K. Whitcraft is the Deputy Director and Program Manager for NASA Harvest, a diverse Consortium of more than 50 institutions focused on advancing the use of satellite data by agricultural and food security decision makers. She is an Associate Research Professor in the Department of Geographical Sciences at the University of Maryland, and since 2015, she has served as Program Scientist for G20's Group on Earth Observations Global Agricultural Monitoring (GEOGLAM). She serves as Agriculture Point of Contact to the world's space agencies (through CEOS), co-leads GEOGLAM's Capacity Development Team, and is Founder and Director of the Agricultural Monitoring in the Americas Initiative. She is an expert in organizational change with respect to integrating new satellite technologies into work flows. She has developed collaborations and partnership models with public and private sector, emphasizing sustainable business models and value to all actors. Dr. Whitcraft, having grown up working in her family winery, also understands well the challenges of high-quality agricultural production in the context of climate change, extreme weather events, and land mismanagement.

Gabrielle Wong-Parodi is an Assistant Professor in the Department of Earth System Science and Center Fellow at the Stanford Woods Institute for the Environment at Stanford University. Her research focuses on applying behavioral decision research methods to address challenges associated with global environmental change. Dr. Wong-Parodi seeks to understand the psychosocial and contextual factors that influence people's responses to environmental change—especially extremes—over time, with a particular focus on those communities that have been historically marginalized or disproportionately impacted by climate change. She also uses behavioral decision science approaches to create and evaluate evidence-based strategies for informed decision making, with a particular focus on building resilience and promoting sustainability in the face of a changing climate. Dr. Wong-Parodi has a background in climate change adaptation and mitigation, energy technologies and resources, extreme weather events, and low-carbon technologies. She was an invited speaker at the Sackler Colloquia at the National Academy of Sciences on the Science of Science Communication. She recently served on the National Academy of Sciences committee "Long-term Coastal Zone Dynamics: Interactions and Feedbacks between Natural and Human Processes and their Implications for the U.S. Coastline." Dr. Wong-Parodi is an adjunct assistant professor in the Department of Engineering and Public Policy at Carnegie Mellon University. She received her B.S. in Psychology at the University of

Appendix C

California Berkeley, and her M.A. and Ph.D. in Risk Perceptions and Communication from the University of California, Berkeley.

Brian L. Zuckerman is a Research Staff Member at the Institute for Defense Analyses Science and Technology Policy Institute (STPI). Dr. Zuckerman's areas of emphasis at STPI are in the areas of program evaluation and scientometrics, where his work focuses on federal research and development program performance and agency-wide research portfolios. Dr. Zuckerman has also analyzed federal research and development data systems and statistical data collection programs. Before joining STPI, he was a principal at C-STPS, LLC and at the Center for Science and Technology Policy of Abt Associates, Inc. He is a former co-chair of the Research, Technology, and Development Topical Interest Group of the American Evaluation Association. Dr. Zuckerman holds a B.A. in chemistry from Harvard College and a Ph.D. in technology, management, and policy from the Massachusetts Institute of Technology.

Appendix D
Line-by-Line Comments

EXECUTIVE SUMMARY

Page/line	Comment
P02/L1	The approach of the summary seems odd. This reads more as an Introduction than as a summary. The reader does not learn much about what is in the Decadal Plan except for the last paragraph and the four pillars.
P02/L1	The summary could be more aspirational and articulate what is that the Nation will get for this 3-4 $B in the next 10 years
P02/L1	This summary focuses on climate change with add-ons for other global change drivers. I suggest that DSP should look at all drivers on an equal basis, including climate change. The DSP should recognize the different scale at which different drivers operate. Climate change will have large impacts in decades to centuries but land-use change, soil erosion and biodiversity loss may have negative impacts now.
P02/L10-11	One thought is to begin the Executive Summary by discussing global change issues and noting that climate change is probably the most consequential but other aspects of global change such as human effects of biodiversity, land use, and (I suggest adding) plastics (or persistent pollutants) also have significant impacts on human and natural systems.
P02/L23	Some care needs to be taken in how these other changes are listed (e.g. why only one kind of pollution). Could also align this list with the headings on pp 9-16 (pillar 1).
P02/L25-26	Can we say which "other factors are often disproportionately affected"?
P02/L34	Add "net" so the sentence reads "transformational efforts to reduce net greenhouse gas emissions".
P02/L39-40	The sentence "New knowledge and approaches are needed to inform measures to adapt..." is correct. It could be made sharper and more timely but noting that with the passage of the Bipartisan Infrastructure Bill, the US has significantly increased its investments in infrastructure and it is critical that such investments adequately account for future climate change risks.
P02/L40	Be more specific regarding what new knowledge and approaches i.e. social sciences, science and technology studies, the humanities etc...
P02/L40	Suggest replacing "projected" in the phrase "adapt to current and projected impacts" with "future" because we're dealing with many impacts that had not been projected.

Page/line	Comment
P02/L40	Replace "carbon" with "net greenhouse gas" in the phrase "reduce carbon emissions".
P02/L43	For the non-climate world "mitigation" has a somewhat different meaning. Add "Climate" at the start of reference box 4.
P03/L1	Be more specific when referring to "social groups", i.e. at the intersection of gender, race, class, sexuality etc...
P03/L2	Perhaps also consider understanding where adaptation/mitigation is working well in impacted communities to see what we can learn, and how this can be rapidly translated to other contexts.
P03/L4	Add "communications" into the list of essential systems the support society stated here.
P03/L8-11	There is also a need to act with the information that we have already; how much information is "good enough"...
P03/L9	Is the research on disruptions also translational research on how to improve preparedness & response?
P04/L1	The pull quotes are fine but they could be removed to save space, if need be.
P04/L17	Funny font switch in "eutrophication of Earth's ecosystem".

INTRODUCTION

Page/line	Comment
P05/L14	Climate is part of global change.
P05/L30	Add NASA to caption for agency logos.
P06/L10	Add social, psychological, and environmental as impacts alongside "economic costs".
P06/L14	Highlight that while the burden is "felt equally across society", it was not caused equally across society.
P06/L17	Rephrase to include "people of color" in this example.
P06/L17	Highlight that redlining is primarily an issue of racism, but race is not mentioned in this sentence or subsequent sentences.
P06/L19	Highlight these neighborhoods face "higher risks of death from heat- related impacts" because they have less green spaces.
P06/L19	Highlight that the issue of race is important to include, women are also more vulnerable to climate hazards.
P06/L21	In science, policy and action can exacerbate inequalities, this is also an issue of doing work under structures that are discriminatory to women and people of color.

Appendix D

Page/line	Comment
P06/L23-33	Box 3: Consider using the easy 4-part descriptions here and their relevance to enviro/climate change: McDermott M, Mahanty S, Schrekenberg K. 2013. Examining equity: A multidimensional framework for assessing equity in payments for ecosystem services. Environmental Science and Policy 33: 416–427. Pascual, U., Phelps, J., Garmendia, E., Brown, K., Corbera, E., Martin, A., ... & Muradian, R. (2014). Social equity matters in payments for ecosystem services. Bioscience, 64(11), 1027-1036. Re: protected classes (which the federal agencies should be including ostensibly in their research), it should be sufficient to include a reference to the 8-11 defined classes that agencies should classify for relevance to their medium or statutory authority, and include some reasonable "above class" guidance, like the Justice40 definitions of underserved communities. [https://www.whitehouse.gov/wp-content/uploads/2021/07/M-21-28.pdf]
P06/L29	"Diversity of thought, knowledge, and experience" can be simplified to "different worldviews".
P06/L30	Include "practices" after "and responses from the perspectives".
P06/L30	Equity and environmental justice need definition. The EPA has a nice definition for EJ. We can use a UN definition for equity.
P06/L32	Have citations for why "build capacity and a more diverse scientific workforce" this is important.
P06/L33	Define "frontline communities".
P06/L37	Change "to play" to "as a coordinator".
P06/L38-39	As climate change science and policy is seeking to widen its base, it is also important to define what roles new actors play or should play. Recognize and provide guidance on the role of groups such as Businesses, NGOs, Civil Society, States, Municipalities, households etc.
P07/L5	Specify if it's "the risk the systems pose" or "the risk the interactions pose". It's important to notice that these systems also provide enormous benefit in the context of global change as well.
P07/L19	Having this and the glossary seem somewhat redundant. If looking for more space, this is a place where text could be reduced.
P07/L21	Use the IPCC definition of climate change, which includes natural as well as anthropogenic change.
P07/L36-37	It would be helpful to have some examples such as the definition of "global change" here.
P08/L1	This definition excludes resilience of natural systems, which cannot anticipate or prepare.

ADVANCING SCIENCE

Page/line	Comment
P09/L2	The way the strategic plan is described here and throughout the document feels as if this document is not the plan itself. Talking about "this plan" and having it be more present might make it also feel less abstract/more urgent and more engaging for those who implement it.
P09/L2	It's not clear what "systems-based research" is.
P09/L15-19	Perhaps mention the USGCRP Advisory Committee?
P09/L19-21	Many of these acronyms have already been introduced and can be used here.
P09/L29	Cascade?
P09/L30	Change "manage" to either address, adapt, or mitigate in the phrase "user-driven science needed to manage climate and global change."
P09/L34	Another bullet needs to be added here, something that goes beyond quantifying, especially if it's human systems (e.g., supporting research networks, providing a framework, ...).
P09/L38	Include "rivers and streams" in the list of "in-water" examples.
P10/L8	Add "and nature." after "pose critical risks to society".
P10/L11	And actions to address?
P10/L11	Reframe this header to "Advance understanding of the physical and social processes that drive the nature and outcomes of extreme events." That would open up research examples on social interventions to reduce risk.
P10/L11	Bigger related idea: The societal impact part of the "Advancing Science Section" thing might need to go up front & be mentioned as connected to the subsequent areas of advancement, because extreme events, tipping points, biodiversity, land use are all issues that disproportionately impact marginalized communities.
P10/L36	Add "or the natural world" after "significant impacts on society".
P11/L2	Good to see this section. Should mention other key drivers of biodiversity reduction e.g., land use, pollution.
P11/L12	This clause does not parse, maybe delete "due" in "losses due in marine"?
P11/L28	Globalization, fuel crises, and changing consumption patterns are important processes that could be identified in this list.
P12/L2	Climate Sensitivity and feedbacks should not be a priority. It is a very important scientific issue, but given how much research has been devoted to this topic, and that the AR6 just narrowed the uncertainty range for climate sensitivity, one has to wonder how much gain the USGCRP expects to make on this matter versus other topics. For example, could more resources be put into better understanding drivers of modes of climate variability and their relationship with anthropogenic driven climate change?
P12/L12	"observational networks" are foundationally important—in the physical and social science.
P12/L28	Consider phrasing: scientific discussions of uncertainty in models can be misunderstood as questions about the underlying science.

Appendix D

Page/line	Comment
P12/L32	Climate sensitivities and uncertainty investigations should heavily be guided by user input. Where and how does climate sensitivity and uncertainty matter in decision-making?
P12/L38	This section is written as "expand research" implying, "do more of what we're already doing" rather than "do wholly new things". The "fulfilling the vision" section suggests this section involves new priorities and activities for USGCRP. Would be worth calling those out..... Also don't see reference to "user-driven" science here.
P12/L43-44	Could broaden this by instead using "human security" or just "security" instead of "national and international security." That encompasses national and international security, but also domestic tranquility, violence. There is published research projecting increases in crime caused by climate change.
P13/L11-13	Are there other approaches besides modeling that can produce insight into coupled human-natural systems? It seems that social science has applied more than modeling to develop such insights.
P13/L11-13	Provide some suggestions regarding different methodological approaches and disciplines—ethnography, photo-voice, interviews, surveys, etc.
P13/L17-18	Change "fully integrating" to "integrating" or "better integrating" in the phrase "human-natural systems requires fully integrating the social and natural sciences" as "fully integrating might limit or slow progress.
P14/L4	What about co-solving for climate challenges in the context of other issues at the center of community life? e.g. food security, environmental health, racial equity in city planning?
P14/L8	Other considerations such as equity should be applied not just "economic based models to evaluate societal decisions."
P14/L16	"co-design, co-production, and co-dissemination" all fall under the umbrella of co-production. The Lopez reference is a case study from the Jordan River - if these references are intended to be pointers to information to help people implement the plan, we might consider recommending some others.
P14/L16	May need to caveat that coproduction must proceed carefully, such that it does not demand an inordinate amount of affected communities and properly compensates time and efforts. Moreover, that the process does not reify existing imbalanced power structures.
P14/L20	Unexpected?
P15/L2	Power, geography, knowledge, technology, structures and institutions all exacerbate vulnerability beyond environmental hazards.
P15/L2	One of the things that resonated about the 2009 NAS report on Earth's Energy future was that at the end of the list of recommendations there was a statement "a number of current barriers are likely to delay or even prevent the accelerated deployment of the energy-supply and end-use technologies described in this report. Policy and regulatory actions, as well as other incentives, will be required to overcome these barriers." What can be said about navigating local to international structural barriers to change? This could be a good place to specify equity too. This section kind of reads like the vulnerability is to all equally.

Page/line	Comment
P15/L8-12	Suggest dropping the example from Zinsstag et al., 2018.
P15/L14	Here and in general, I think it's helpful that this research will be conducted, but I wonder how these efforts might be coordinated to truly move the ball effectively down the field. The strategic plan could provide an opportunity to strategize about what coordination might look like... (would welcome more consideration on that throughout this doc).
P15/L26	Social sciences is an academic field, the other topics in the list are elements of the climate system.
P15/L27	Insights and methods?
P15/L32	Add "practices that support" after "effective responses need to incorporate considerations."
P15/L35	This is a good list of research topics. If space permits it might be interesting to explore how these could be applied to specific global change research topics. For example, how would these be applied to examine loss of biodiversity?
P15/L36	I know this term "human systems" has been used above but I wonder if it is effective...
P15/L39	Valuing?
P15/L41	Could be implementing too and not just valuing.
P15/L42	"conditions of deep uncertainty" needs to be challenged.
P16/L5-7	Should include other global change responses e.g., how reduce biodiversity loss, plastic pollution.
P16/L9	Humanities?
P16/L12	There is a broad literature on "just transitions" that speak about the importance of justice and equity in such shifts.
P16/L15	Add "costs" to "the benefits, trade-offs, path dependencies, and interactions".

INFORMING DECISIONS

Page/line	Comment
P17/L1	Social science and humanities are missing here. How do we ensure that science is operationalized into policy and action?
P17/L2	This is a very good discussion, but it only covers climate change decisions.
P17/L29-37	Strong/defined actions, "will provide/ is ready to inform" implies sufficient certainty for decision-making – same question about how this pillar can be met given state of the science/speed of advancement of the science.
P17/L36-37	This is where equity must be considered to make sure analyses doesn't miss, see recent example from flood plain analyses: https://journals.sagepub.com/doi/full/10.1177/2378023120905439.
P17/L36-37	It's the distribution of benefits and costs that is very important here (not just the overall costs and benefits).
P18/L2-3	How do we address lack of capacity to act? There might be enough information but are there enough resources to act on this information?
P18/L2-3	This also includes incentives.

Appendix D

Page/line	Comment
P18/L3	With respect to capacity there is the capacity to act, as well as the capacity to engage.
P18/L4	It is important to improve and sustain existing tools, as well as build new ones. And, where possible incorporate information into the tools decision makers already use.
P18/L4-5	This is very good. It's also an example of a specific goal that there should be more of in the DSP. Any sense of where we will be in 10 years on this?
P18/L12-13	Yes, and this information can feed back to help determine important future research questions.
P18/L17	Why do assessments fit within informing decisions? Assessments should be part of the scientific process and engaging the Nation.
P18/L20	Section 2921 is the definitions. NCA is section 2936 (15 U.S. Code § 2936 - Scientific assessment	U.S. Code	US Law	LII / Legal Information Institute (cornell.edu))
P18/L25	This is big. I imagine they didn't have time to add it to other places in the document yet, but should (e.g., biodiversity section).
P19/L19	What are equity dimensions?
P19/L23-34	Can be merged with Pillar 3 discussion of research design on page 23—if purpose is "engagement" then belongs in Pillar 3; if purpose is "useful information" then belongs in Pillar 2.
P19/L24	Yes! Kudos to USGCRP and the DSP authors for moving in this very important direction.
P19/L26-27	Yes, and it would be great to see the USGCRP doing some coordination with these partnerships to reach more people (add avoid duplication and stakeholder fatigue) and to help make partnerships as effective as possible (a "no wrong door" policy—if one agency gets involved and then realizes their partner needs something they can't provide, there's a way to connect them to a more aligned partner).
P19/L31-34	This is very important. Can the discussion be longer to get into some more specifics?
P19/L32-33	There could be a place to share translation models, because it isn't clear what this means.
P19/L36	"Indigenous and Traditional Ecological Knowledge" would be better in the section about advancing knowledge.
P20/L1	Will consider?

ENGAGING THE NATION

Page/line	Comment
P21/L12-13	Add: "or have the capacity to act or engage."
P21/L23-24	Very good. To what extent do USGCRP members engage with information consumers within their departments and agencies?

Page/line	Comment
P21/L29	People will need to be more than connected to tools, they will need training on how to use them & at what scales & could also be encouraged to bring others into tool use.
P21/L31-33	Explicitly defining the CASCs/Climate Hubs as being outside of the USGCRP (despite being funded by agencies that participate in USGCRP) is noteworthy.
P21/L37	It's called the Cooperative Extension System now and is part of NIFA (Cooperative Extension System | National Institute of Food and Agriculture).
P21/L40	A strategic plan of this magnitude should not mention something as mundane and small-scale as webinars.
P21-22/L40-L2	This mini-paragraph appears disconnected without some proper nouns—the verbs are all passive and give no sense of who will do these things and how they connect to the strategic plan. Is the point that the federal regional science organizations will do this as part of/in concert with USGCRP? Consider rewriting or dropping.
P22/L4-8	This statement is broader than the first paragraph of the regional science organizations statement – no proper nouns. Perhaps some examples of kinds of boundary organizations (e.g., scientific/professional societies, perhaps)? This would be a good place to discuss geography/rural areas and engagement with faith communities.
P22/L7-8	Add: "and the feedback from decision-making to research"
P22/L12-13	The difference between "opportunities to interact" and "engagement" is cryptic. If the point is that "engagement" = "FACA" then this sentence belongs at the end of the previous paragraph.
P22/L16	If "engage" = "FACA" in the previous section, is this the verb that should be used here?
P22/L19-20	"integrate the organizations" reads as "have them merge"—is that what is meant?
P22/L21-23	Good. Perhaps add community based organizations.
P22/L25-29	This paragraph isn't about "engagement" per se. Arguably it's about Pillar 2. Needs some tightening.
P22/L32	What is the education for this workforce? Is professional development to build skills & climate leaders, part of this or just K-12, informal ed, & extension/community engagement. Describe multi-level more.
P22/L33	Add a note about the difficulty of getting a diverse workforce when not all social identity groups and different socioeconomic backgrounds get to places where they can be seen and picked.

Appendix D

Page/line	Comment
P22/L36	If this is true, then the mix of university types (land-grant, HSI, MSI, tribal) isn't sufficient. Need language around building global change research at community/two-year colleges and state/local university systems (not just the flagship land-grant institutions).
P22/L40-44	Point about preferential involvement of HBCUs/MSIs from previous section might go here.
P22-23 /L40-2	What is translation? Is it education (k-gray), professional development (e.g. supporting climate professionals)? Training to do what?
P23/L5-6	Is it also possible to add something here about using education to help translate education, we need a systems approach not just more science with community members, but once that science is done continue with education, not everyone has access to research participation, but what education looks like is HUGE (e.g. are K-12 standards support the literacy & agency/belonging needed for equitable action).
P23/L7-8	This is important, but it's not "research design".
P23/L8-10	"Applied science" is mentioned neither in Pillar 1 nor Pillar 2. Which parts of the Pillar 1 research program are "applied" science? How does "community-level adaptation" fit in either to "applied science" or to decision-making?
P23/L12-13	Agree these are critical. "Rapid assessments" are only mentioned here, and "synthesis" is only mentioned in the context of biodiversity (Pillar 1). Who is carrying them out? Are they Pillar 1 activities? Pillar 2 activities?

COLLABORATING INTERNATIONALLY

Page/line	Comment
P24/L4-5	Arguably "responding" isn't inherently international (though responding effectively probably is)...
P24/L11-12	Collaboration across boundaries is not necessarily related to collaboration across disciplines, but the language carries throughout the Pillar.
P24/L14-17	Not sure this statement is the right tone. "is fostered by" might be better...
P24/L19	Drop the acronym here.
P24/L29	This flagship is missing a verb—other three have verbs.
P24/L34	References refer to 1987, International Geosphere-Biosphere Programme
P24/L37	IPCC acronym defined earlier.
P25/L6	Here would be a good entry into increasing diversity in climate science.
P25/L13	NCA acronym already introduced.

Page/line	Comment
P25/L19	Foster instead of enhance?
P25/L35	Define acronym WCRP here?
P26/L7	"Building capacity" is parallel.
P26/L40-41	Key to bring in social sciences and humanities to better address the inability of scientific knowledge to be translated into action via policy.
P27/L9-10	Conversations between scientists and policy-makers are essential here as much of the scientific reports, as rigorous as they can be made to be, still, do not carry much weight at the UNFCCC.
P27/L11	They are technically "intergovernmental" GEO.
P27/L11	Needs acronym defined here (USGEO United States Group on Earth Observations).

FULFILLING THE VISION

Page/line	Comment
P29/L1-2	Introduce this (particularly what opportunities could be), earlier in the document.
P29/L1-4	Evaluation in the rest of the document either refers to science activities or evaluation of the USGCRP's products and outputs. This is a stronger statement about assessing US government capacity and policy.
P29/L9	Sexual orientation is only mentioned here—include elsewhere where minority groups are listed.

GLOSSARY

Page/line	Comment
P30/L15	May not want to put "weather" in the definition of climate change.
P30/L16	Climate change is not necessarily anthropogenic.
P30/L21	This framing perpetuates the "communication to" / loading dock model that the DSP mostly moves beyond. Update text to make multi-directional and affirm that both parties are knowledge producers.
P30/L30	Date in Lubcheco letter is 18 May 2021.
P30/L33-34	Edit definition. In and out of atmosphere is one part of it, but have also seen to and from surface of the earth (e.g., An update on Earth's energy balance in light of the latest global observations | Nature Geoscience).
P30/L43	Only mentioned here. Consider incorporating religious communities into main text?

Page/line	Comment
P31/L1	Only mentioned here. Consider incorporating persons with disabilities into main text?
P31/L1	"geography" in the conclusion, "rural" not mentioned explicitly anywhere else.
P32/L4-5	This is very specific to get a glossary mention (discussed only once in the document).
P32/L7-9	This is very specific to get a glossary mention (discussed only once in the document).